Gurdon Buck

Contributions to Reparative Surgery

showing its application to the treatment of deformities, produced by destructive

disease or injury - congenital defects from arrest or excess of development

Gurdon Buck

Contributions to Reparative Surgery
showing its application to the treatment of deformities, produced by destructive disease or injury - congenital defects from arrest or excess of development

ISBN/EAN: 9783337239572

Printed in Europe, USA, Canada, Australia, Japan

Cover: Foto ©berggeist007 / pixelio.de

More available books at **www.hansebooks.com**

CONTRIBUTIONS

TO

REPARATIVE SURGERY:

SHOWING

*ITS APPLICATION TO THE TREATMENT OF DEFORM-
ITIES, PRODUCED BY DESTRUCTIVE DISEASE
OR INJURY; CONGENITAL DEFECTS FROM
ARREST OR EXCESS OF DEVELOP-
MENT; AND CICATRICIAL CON-
TRACTIONS FROM BURNS.*

BY

GURDON BUCK, M. D.

ILLUSTRATED BY NUMEROUS ENGRAVINGS.

NEW YORK:
D. APPLETON AND COMPANY,
549 AND 551 BROADWAY.
1876.

PREFACE.

THE volume now offered to the medical profession contains the author's own experience in the department of practical surgery of which it treats. While it makes no claim to be a systematic and thorough treatise on the subject, it will, nevertheless, be found to contain, in a classified order, much that is treated in existing works on plastic surgery. It is believed also, that, in addition to this, something new and useful has been contributed to the resources of surgical art. There is no department of surgery where the ingenuity and skill of the surgeon are more severely taxed than when required to repair the damage sustained by the loss of parts, or to remove the disfigurement produced by destructive disease or violence, or to remedy the deformities of congenital malformation. The results obtained in such cases within the last half century are among the most satisfactory achievements of modern surgery. The term "Reparative Surgery," chosen as the title of this volume, though it may, in a comprehensive sense, be applied to the treatment of a great variety of lesions to which the body is liable, is, however, restricted in this work exclusively to what has fallen under the author's own observation, and been subjected to the test of experience in his own practice. It largely embraces the treatment of lesions of the face, a region in which plastic surgery finds its most frequent and important application. Another and no less important class of lesions will also be found to have occupied a large share of the author's attention, viz., cicatricial contractions following burns. While these cases have a very strong claim upon our commiseration, and should stimulate us, as surgeons, to the greatest efforts for their relief, they have too often in the past been dismissed as hopelessly incurable. The satisfactory results obtained in the cases reported in this volume will encourage other surgeons, we trust, to resort with greater hopefulness in the future to operative interference. Accuracy of description and clearness of statement have been aimed at in the following pages; and if, in his endeavor to attain this impor-

tant end, the author has incurred the reproach of
tediousness, the difficulty of the task must be his
apology.

The indispensable aid of pictorial illustration
has been employed to the fullest possible extent;
and whatever success has been achieved in this
department, the credit of it is due to the admi-
rable skill of Ferdinand Froning, of Vienna, who
executed all the principal figures from photographs.
Special pains have also been taken to ascertain the
permanent results of all the cases reported, after the
lapse of long intervals of time, and fortunately this
has been practicable in almost all instances.

Two albums, containing each a collection of the
photographs from which the portrait figures illustrating this volume were engraved, have been deposited, the one in the Pathological Museum of the
New York Hospital, the other in the United States
Army Medical Museum, at Washington, D. C., where
they are accessible to any person desirous of con-
sulting them. Each photograph is designated by
a number corresponding to the figure in the volume
of which it is the original.

Of the cases reported in this volume, the following have previously appeared in print:

CASE I. In "Transactions of the Medical Society of the State
            of New York," vol. 1864.
      II. "Transactions of the American Medical Association,"
            vol. 1870.
     IV. *American Practitioner*, August, 1873.
     VI. *American Journal of Medical Science*, vol. lviii., new
            series, 1869, p. 352.
   VIII. "Transactions of the Medical Society of the State
            of New York," vol. 1866.
     IX. *New York Medical Record*, February 1, 1871.
      X. *Boston Medical and Surgical Journal*, January,
            1874.
     XI. XII., XIII., XIV. *New York Medical Record*, January 15, 1872.
   XVII. *New York Medical Record*, 1866.
    XXI. *New York Medical Record*, 1870, p. 483.
   XXII. "Transactions of the New York State Medical Society," 1872.
    XXV. *New York Medical Record*, 1869.
  XXVII. *American Journal of Medical Science*, 1872, p. 52.
 XXVIII. *New York Medical Record*, 1874, p. 473.

CONTENTS.

CHAP.   PAGE
 I. TRANSPLANTATION OF SKIN   . 7
 II. METHODS OF TRANSFER 10
 III. TREATMENT OF RAW SURFACES LEFT TO HEAL BY GRANULATION . 13
 IV. SUTURES AND THEIR MANAGEMENT 14
 V. METHODS OF OPERATION. 20

CASES.

FIRST CLASS.—LOSS OF PARTS INVOLVING THE FACE, AND RESULTING FROM DE-
STRUCTIVE DISEASE OR INJURY.

CASE I. Reconstruction of the Mouth and Repair of the Nose after the Loss
of the Right Half of the Upper Lip, the Adjacent Portion of the
Cheek and Ala Nasi, together with the Entire Right Superior
Maxillary Bone 31

 II. Reconstruction of the Mouth and Repair of the Nose after the Loss
of the Right Half of the Upper Lip, and Adjacent Portion of the
Cheek, and Right Ala Nasi 52

 III. Reconstruction of the Nose and Mouth after the Loss of the Nose
and Entire Upper Lip 62

 IV. Reconstruction of the Mouth after the Loss of the Entire Under
Lip, and a Portion of the Inferior Maxillary Bone . . 75

 V. Reconstruction of the Mouth after Removal of the Under Lip for
Disease 87

 VI. Reconstruction of the Mouth after the Loss of the Right Half of
Both Lips; also, Esmarch's Operation for Anchylosis of the Jaw 93

 VII. Mutilation and Distortion of the Mouth from a Shell-Wound . . 101

 VIII. Loss of a Large Portion of the Lower Jaw-bone, with Extensive
Mutilation of the Face and Distortion of the Mouth, produced
by a Shell-Wound 108

 IX. Closure of an Opening into the Superior Meatus of the Right Nasal
Fossa 114

 X. A Rhinoplastic Operation for the Restoration of the Apex Nasi
after it had been bitten off 118

SECOND CLASS.—CONGENITAL DEFECTS FROM ARREST OR EXCESS OF DEVELOPMENT.
—HARELIP.

PAGE

CASE XI. Single Harelip 131

XII. Double Harelip 132

XIII. Single Harelip, with Cleft of the Dental Arch and Bony and Soft .
Palates 134

XIV. Single Harelip on the Right Side 137

XV. Single Harelip 138

XVI. Single Harelip 139

XVII. Double Harelip, complicated with Cleft of the Bony and Soft
Palates, together with the Presence of an Intermaxillary Bone 140

XVIII., XIX., XX. Examples of Cases of Harelip, operated on a Second
Time for the Purpose of remedying the Imperfect Results of
Previous Operations performed in Infancy . . . 144–146

XXI. Congenital Hypertrophy of the Tongue 151

XXII. Congenital Hypertrophy of the Under Lip 159

XXIII. Abnormal Growth of Hair upon the Forehead . . . 165

XXIV. Erectile Tumor of Large Size 168

XXV. A Pendulous Tumor (Molluscum Fibrosum) arising from the Right
Half of the Forehead and Temple 172

THIRD CLASS.—CICATRICIAL CONTRACTIONS FOLLOWING BURNS.

XXVI. Disfigurement of the Face from Cicatricial Contractions ; Extirpation of one Eyeball; Closure of the Orbit by a Plastic Operation 185

XXVII. Cicatricial Contractions involving the Chin and Front of the Neck 197

XXVIII. Cicatricial Contractions involving the Face and Hand . . . 206

XXIX. Cicatricial Contractions involving the Right Axilla and Arm 228

REPARATIVE SURGERY.

To avoid repetition in the succeeding narrative of cases, some general observations on the subjects designated by the titles of the following chapters will now be presented.

CHAPTER I.

TRANSPLANTATION OF SKIN.

Choice of Material.—In the choice of skin, which is the material employed to supply an existing deficiency, care should be taken that it be in a normal and healthy condition. If, in consequence of a previous burn, the surface has become glossy, pale, and of cicatricial formation, a patch taken from it for transplantation, by dissecting it up from its underlying connections, would inevitably slough. Such a patch could not maintain its vitality if nourished by the circulation it would receive through its pedicle alone; the vascular support derived from its connec-

tion with underlying parts could not be dispensed
with. The author's experience on this point conclu-
sively establishes the fact that cicatricial integument
cannot be relied upon for transplantation, and should
not be used for that purpose. Another condition to
be observed is, that a patch of skin intended for
transfer should have its long axis correspond to the
direction in which the arterial vessels are distributed,
and the free extremity of the patch should point toward their destination. Upon the forehead, for example, these important conditions are best secured by
raising the patch of skin required from above the
inner half of the eyebrow, where it would retain its
connection and derive its vascular support from a
branch of the ophthalmic artery, which emerges from
the orbit through the supra-orbitar notch.

Precision in Adaptation.—To secure precision in
adapting a patch of skin to a new locality to which
it is to be transferred, the following method will be
found satisfactory. After having prepared the space
which it is intended to fill up, by paring its edges,
and dissecting them up sufficiently from their underlying connections to permit them to be everted, and
also, if necessary, by notching the edges at intervals
to facilitate their eversion, an exact pattern of the
space should be cut from oiled silk, and the pattern
applied to the surface which is to supply the new
material. Small pins may then be temporarily in

serted erect in the skin at intervals around the pattern, and at a distance of one line from its margin,
as an allowance for shrinkage. A larger allowance,
however, must be made for the length of the patch
of skin beyond that of the pattern itself, so as to
permit the patch to be brought around edgewise
to its new location, without causing any strain at
its pedicle, which might obstruct the circulation,
and thereby endanger the vitality of the patch.
The outline of the pattern having been indicated
by the pins, the pattern itself may be dispensed
with, and the pins alone left to serve as a guide.
After the incision defining the patch of skin has
been made, the patch itself should be dissected up
from its underlying connections. Special care should
be taken on the forehead not to wound or detach the
pericranium.

CHAPTER II.

THE transfer of a portion of skin to a neighboring locality may be made by different methods, the choice of them depending upon the condition of the parts involved. In the cases hereinafter reported the methods employed were the following:

First Method. By Approximation.—If on both sides of a space requiring to be filled up the adjoining skin is supple and movable, the opposite edges of the space may be pared, and the adjacent skin dissected up from its underlying connections to a sufficient distance to permit the edges to meet and be secured in contact by sutures. If there be any strain on the sutures after the new adjustment, it should be relieved by an incision through the skin on either side of the wound, parallel with, and at a suitable distance from it.

Second Method. By Sliding.—If, upon one side only of a space requiring to be covered, the skin is in a condition to be made available for transfer, a patch of the required size, adjacent to the space, may then be dissected up, and, being left connected at one end, may be glided edgewise toward the space, and adjusted to it by sutures; the edges of the space must of course have been prepared beforehand for the purpose. The surface still left bare by the transferred patch may sometimes be covered by dis-secting up the neighboring skin, and stretching it across the bare surface till it meets the edge of the patch, where it may be adjusted by sutures.

Third Method. By Transfer to a Distance.— Sometimes the material for supplying a deficiency has to be taken from a locality at a distance from the spot where the deficiency exists. This may be done in two ways: 1. A patch of skin, after having been dissected up from its underlying connections, but still remaining attached at one extremity, may be transferred edgewise, and made to describe in its circuit a quadrant of a circle, or even an entire semi-circle, in reaching its new destination. In order that the raw under-surface of the patch, when thus trans-ferred, may lie in contact throughout its entire length with subjacent raw surface, the intervening skin must be displaced. This displacement, however, should be effected in such a way that the displaced skin, retaining a connecting pedicle for its support, may be made to change places with the transplanted patch, and thus contribute, as far as it can, a covering for the surface that has been left bare. This method has been advantageously employed in several of the cases reported in this volume. In all of them the forehead was the seat of operation, and in no instance has a patch of skin, taken from the forehead in the manner already described, failed to do well from deficient vitality, or sustained any loss by sloughing. 2. The transfer of a patch of skin to a distant spot may be effected by a jumping process, as follows:. An elongated patch of skin having been dissected up, but left attached at one extremity, may be made to jump over an intervening sound surface, and have its free extremity adjusted to a distant spot where a deficiency exists. After becoming ingrafted in its new locality, the patch is to be severed, and its pedicle replaced upon the bare surface from which it was originally taken. But one example of this method occurs in the following pages (Case III., p. 62).

CHAPTER III.

AFTER the transfer of a patch of skin to a new
locality, more or less of the surface from which it
was taken must be left to heal by granulation. The
following method of treating such a surface has proved
uniformly satisfactory, and is therefore confidently
recommended. After all hæmorrhage has ceased, the
raw surface should be coated first with a uniform
layer of scraped lint, and then with an additional
layer of lint saturated with collodion. This dressing soon stiffens and forms an artificial scab, which
will remain adherent for from six to ten days, when
it becomes detached by suppuration. In the mean
time the wound requires no other application; it remains in a quiescent state, with scarcely any surrounding redness or inflammatory tumefaction, and only a
slight discharge of matter escapes from one or more
points at the margin of the crust. On its separation
a growth of healthy granulations will be found at
the margin of the sore, and sometimes covering its entire surface, even up to the level of the surrounding
skin. The application just described will hereafter
be known and referred to as the collodion crust.

CHAPTER IV.

In plastic operations success depends so much on the proper management of sutures, that a particular description of the different methods of employing them will not be out of place. Three kinds of sutures have been employed by the author.

1. The Interrupted Thread Suture.—For the insertion of sutures of this sort, trocar-pointed needles, such as are used by glove-makers, are preferable to any other. They may be had of all sizes at the thread-and-needle stores. To guard against inversion of the edges of a wound, which is a frequent cause of the failure of primary union, the needle should be inserted in such a manner that it will pass obliquely through the thickness of the skin, and in doing so the deepest part of its track will be farther from the confronted edges of the wound than it is at the points of entrance and exit upon the surface. Thus inserted, the thread on being tightened tends to evert the edges of the wound, and bring their confronted cut surfaces more perfectly into contact. The confronting of the edges is further facilitated if, after inserting the first suture, instead of tying the knot, the ends of the thread are made use of to draw out the edges of the wound, while a second neighboring suture is being inserted and tied; the same may be done with each additional suture in succession. The sutures should also be inserted as closely together as may be necessary to secure exact coaptation of the edges. Their multiplicity is not objectionable, inasmuch as at the expiration of twenty-four hours, when agglutination of the edges will have taken place, their number may be diminished by removing the alternate ones. Metallic sutures, in the judgment of the author, possess no advantage over thread, and are as liable to cause ulceration as thread. Both will remain harmless as long as they are needed, provided the edges of the wound they are designed to hold in contact have previously been liberated, so as to be relieved of all strain upon them.

2. **The Pin, or Figure-of-eight, Suture.**—The insertion of the pin is performed, with facility and precision, by the aid of an instrument devised by the author, and first described in the *New York Medical Record*, of July 1, 1869, under the name of "Suture Pin Conductor." It consists of a needle two and a half inches long, of the thickness of an ordinary knitting-needle, slightly curved toward the point, and fixed in a handle. From its point toward the

2

handle, its thickness grows smaller for a distance
of half an inch, which facilitates its passage through
the skin and beyond. Its extremity is beveled off
to a sharp point on the concave side of the needle,
and is perforated lengthwise for a short distance on
its beveled face, like the point of an hypodermic
syringe. (*See* Fig. 1, *a.*) It is used in the following manner: The edges of the wound which are to
be approximated, having been traversed by the conductor guided by one hand, a pin, held between the
thumb and fingers of the other hand, is engaged by

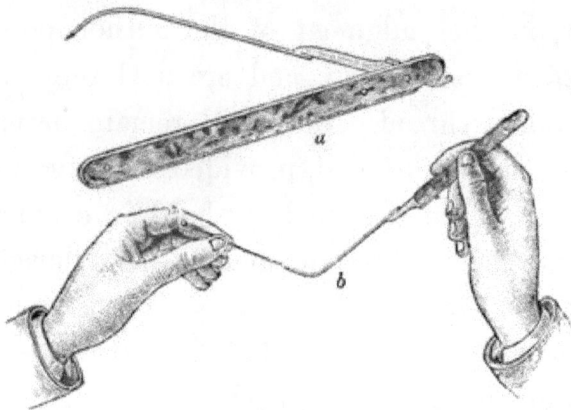

Fig. 1.

its point in the perforated hole at the point of the
conductor, and held steadily in place while the conductor is withdrawn. (Fig. 1, *b.*) The pin follows
it with perfect certainty to its place, and is then
wound with loosely-twisted cotton-yarn passed in
figure-of-eight turns around both ends of the pin.

The advantage of this instrument is, that with it the pin can be inserted with great precision, and fresh pins, when necessary, can be inserted near to and to supply the place of old ones requiring to be removed, without disturbing the newly-formed adhesions. To guard against suppuration in the track of the pin, and on the surface of the skin underlying the yarn, the yarn itself should be removed at the end of forty-eight hours, and sometimes of twenty-four hours, so as to allow the constricted surface to recover itself; after which fresh yarn should be applied. This change should be repeated daily afterward, till the removal of the pin on the fourth day, beyond which time it should not be continued in place. If the support afforded by the pin cannot yet be dispensed with on the fourth day, it is better to insert a fresh pin near by than to leave the old one in place longer than four days. A patch of three or four thicknesses of adhesive plaster stuck together, and cut of a proper shape, may be laid upon the surface of the skin, between the points of entrance and exit of the pin, for the yarn to rest upon, and to serve as an additional protection against ulceration.

3. The Beaded Wire Clamp Suture.—This suture is intended as an auxiliary for the support of other sutures, and may be employed where it is important to relieve them of strain, and thereby increase the

chances of obtaining primary union. The following
is the method of inserting it. An ordinary darning-
needle, two inches or more in length, may serve for
the purpose, and should be threaded with a flexible
silver wire, previously charged at its knotted extremity, first, with a small perforated disk of leather
(such as may be found at any saddler's shop), and
then with a smooth, round glass bead. The needle
thus armed is then made to traverse the edges of
the wound at opposite points, and at a distance of
one inch or more from the edges. A second bead

Fig. 2.

should then be strung upon the other end of the
wire. While the opposite edges of the wound are
crowded toward each other and held in contact, the
free end of the wire should be drawn upon, and the
second bead slid down to its place against the skin,
and secured there by winding the wire three or four
times around the end of a friction match, or other
suitable piece of wood. The wire itself should be
left two inches long, so as to permit the clamp to be
tightened or loosened at pleasure.

The further adjustment of the edges of the wound to each other may then be completed by any required number of interrupted thread sutures. The special advantage of the clamp suture is, that it relieves the other sutures of all strain, and admits of being tightened or slackened, if necessary. It may remain *in situ*, undisturbed, from six to ten days, and any superficial ulceration produced by the pressure of the beads against the skin heals readily after their removal, without leaving any scar. Another advantage is, that, if primary union wholly or partially fails to take place, the edges of the wound are still maintained in contact, while union by the slower process of granulation is going on.

CHAPTER V.

THE observations under this head relate only to the reconstruction of the mouth after the entire loss of either the upper or lower lip, or of a considerable portion of either. The two methods about to be described have been successfully employed in several of the following cases, and, in the results obtained, it is believed these methods possess advantages over other methods in use among surgeons, and they also differ from them in some important respects. To avoid repetition, they will now be described in detail.

First Method.—In order to execute this method, a preliminary operation must be performed to bring the parts into a proper condition for it. For example, the under lip being involved, the removal of the diseased portion may be executed in two ways, the choice of which should be determined by the condition of the parts.

First Preliminary Operation is that for the removal of the diseased part, by including it in a letter-V-shaped patch, as follows: An incision, commencing at a point within half an inch of the angle of the mouth on both sides and dividing the lip border should be carried downward on either side of the morbid growth in converging lines, till both incisions meet under the chin in the median plane. The included V-shaped patch should then be dissected up from the underlying periosteum and removed. The lining mucous membrane of the mouth should then be divided on both sides of the wound, along the line of its reflection from the jaw to the inside of the lip, and the division continued outward as far as is necessary to permit the edges of the V-shaped wound to meet at the symphysis, and there be secured together by sutures.

Second Preliminary Operation is that for the removal of the diseased part by including it in a quadrilateral patch. Two incisions, dividing the lip border at a distance of half an inch from the angles of the mouth, should extend vertically downward on either side of the diseased part, till they both connect with a third transverse incision crossing the lower part of the chin. After the removal of the included quadrilateral patch, the transverse incision should be continued outward, on both sides, to within a finger's breadth of the angles of the jaw, and thence upward a distance of two inches on the sides of the cheeks in lines curving forward. The two cheek-flaps thus formed, both of which are lined with mucous membrane, are then to be dissected up from the jaw, and brought forward edgewise, so as to meet in front at the symphysis, where they are to be secured together by sutures. The surfaces left bare on both cheeks by this transfer may be covered again by dissecting up the skin bounding them posteriorly from its underlying connections, and advancing it so as to meet the edge of the transferred cheek-flap and be adjusted to it by sutures. By both these preliminary operations the mouth is made to assume a shape in which the upper lip is folded upon itself, and overhangs the retracted and shortened under lip. Case V., p. 87, furnishes an example of the application of the second method just described. Fig. 3 illustrates the first method as completed on the cadaver. The letter-V-shaped piece, excised from the under lip, remains attached and suspended below the chin, and the edges of the wound are secured together by sutures. After the parts thus adjusted have completely healed and regained their suppleness, the special operative method by which the mouth is to be restored to its natural shape may be undertaken. It is as follows. In order to insure precision in making the requisite incisions, their course should first be designated by pins temporarily inserted erect in the skin at certain points, as shown by Fig. 3. For example, letters *a a* represent two pins inserted

at one finger's breadth below the under-lip border,
one on either side of the chin, a little to the outside
of the angle of the mouth, and both equidistant
from the median line; *b b* are also two pins inserted,
one on either side, into the upper lip at the margin

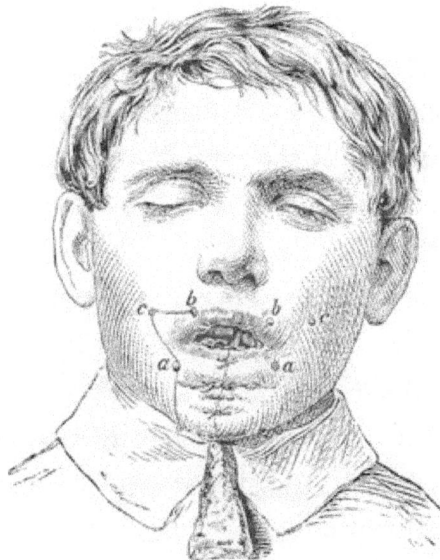

FIG. 3.

of the vermilion border, both equidistant from the
median line, and at such a distance apart as to include between them sufficient length of lip border
with which to form a new upper lip. The steps of
the operation are then the following: With the forefinger of the left hand placed on the inside of the
mouth, the left cheek is to be kept moderately on
the stretch, while with a sharp-pointed "Beer's" cor-

nea knife it is transfixed at the point *a*, marked by the lower pin on the left side of the chin. An incision is then to be carried through the entire thickness of the cheek, upward and a little outward, a distance of one inch and a half, to a point, *c*, near the middle of the cheek. The left half of the upper lip should next be transfixed at the point *b*, marked by a pin on the vermilion border, and the incision carried through the lip and cheek, outward and a little upward, to join the first incision at its terminus *c*, in the middle of the cheek. A triangular patch, *b*, *c*, *a*, will thus be formed, which will include the entire thickness of the cheek, and whose apex will be free and disconnected, while its base remains attached toward the mouth. The next step is to transfer this patch from the cheek to the side of the chin. For this purpose an incision should be made on the side of the chin from *a*, the starting-point of the first incision, vertically downward to the edge of the jaw, and to the depth of the periosteum. The edges of this incision retracting wide apart, afford a V-shaped space for the lodgment of the triangular patch, which is now to be brought around edgewise, and adjusted by sutures in its new locality on the side of the chin. By this transfer the portion of upper-lip border that formed a part of the base of the patch is brought into a transverse line, continuous with the under lip, and constitutes an extension of it. The space upon

the cheek from which the triangular patch was taken
is closed by bringing its edges together, and securing
them in contact by sutures. By this adjustment a
new and naturally-shaped angle is formed for the
mouth at the point b, on the vermilion border, where
the lip was transfixed in commencing the second in-

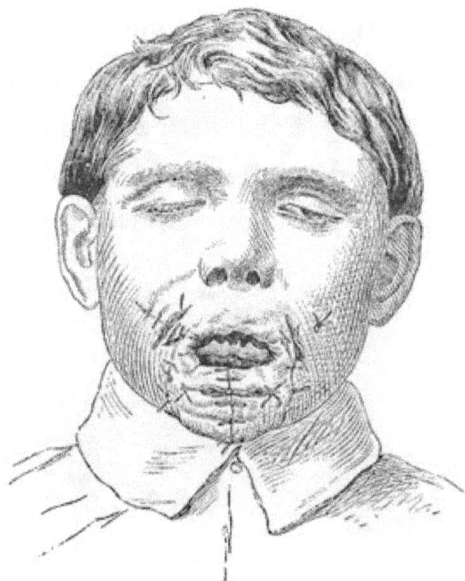

FIG. 4.

cision of the cheek. The incisions described above
should be made with the utmost precision, and special care should be taken that the lining mucous
membrane is divided exactly to the same extent as
the skin itself. The same procedure may be applied
to the other side of the mouth and executed at the
same operation, as was done in Case V., p. 87. The

mouth, when thus reconstructed, is of natural shape
and symmetrical form, as shown on the cadaver and
represented by Fig. 4. It consists of natural lip
border, and is lined throughout with mucous membrane. Orbicular muscular fibres being retained in
the new structure, the natural action of the lips is in
a good degree preserved.

A Second Method. — Another method of recon-
structing the mouth, after the loss of one-half of the
upper lip and an adjacent portion of the cheek, is as
follows: It consists essentially in supplying the deficiency of the upper lip by material taken from the

Fig. 5.

under lip. In such cases the under lip has itself become considerably lengthened transversely. The following are the steps of this operation : the extremity
of the under lip, where it joins the right cheek, is

to be divided through its entire thickness at right angles to its border, and the division carried to the extent of one inch from the border, *a* to *b* (Fig. 5). From the terminus of this first incision, a second incision is to be extended on a line parallel to the lip border, a distance of one inch and a half toward the chin, *b* to *c*. The quadrilateral flap thus formed from the under lip is to be folded edgewise upon itself, and made to meet the remaining half of the upper lip, and be adjusted to it by its free extremity. In order, however, to make this fold, the under-lip flap must first be divided obliquely half across its base, where it still retains its connection with the chin by an incision in the line *c* to *d*.

The left half of the upper lip is also to be prepared for the new adjustment, first, by liberating it so that it can be glided toward the right side. This is accomplished by incising the buccal mucous membrane along the line of its reflection from the jaw to the lip and cheek, and detaching the parts above toward the orbit from the underlying periosteum. Second, by paring a strip of vermilion border from the extremity of the half-lip of sufficient length to permit the end of the half-lip to be matched to the free extremity of the under-lip flap. The parts concerned having been thus prepared, the under-lip flap is next to be doubled edgewise upon itself, and its free extremity adjusted to the half of the upper lip,

and the two secured to each other in a vertical line
below the columna nasi by sutures. The space between the newly-adjusted half of the mouth and the
neighboring cheek is to be closed by approximating
the opposite parts and securing them to each other
by sutures after their edges have been carefully
matched. When the process of healing has been
completed, and the parts have regained their natural
pliability, the newly-constructed half of the mouth

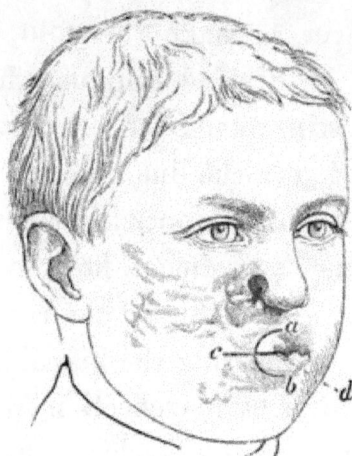

FIG. 6.

—which has assumed a circular form and pouting
shape, as shown by Fig. 6—may be restored to its
natural angular shape and dimensions by a supplementary operation, which may be executed as follows: An incision is to be made with great exactness along the line of the vermilion border circum-
scribing the circular half of the mouth, and extend-

ing to an equal distance on the upper and lower lips
(*a* to *b*). This incision should only divide the skin,
without involving the mucous membrane. A sharp-
pointed, double-edged knife should then be inserted
at the middle of this curved incision, and directed
flatwise toward the cheek, between the skin and
mucous membrane, so as to separate them from each
other as far as the new angle of the mouth requires
to be extended. The skin alone is next to be divided with strong scissors, on a line with the commissure of the mouth outward toward the cheek (*d*
to *c*). The underlying mucous membrane is also to
be divided on a line opposite to, but not so far outward as, the incision made through the skin. The
angle at the terminus of the incision of the mucous
membrane is then to be accurately secured to the
angle at the terminus of the incision of the skin, by
a single thread suture. The fresh-cut edges of skin
and mucous membrane above and below, that are to
form the new lip borders, are to be shaped by paring
first the skin and then the mucous membrane in such
a manner that the latter shall overlap the former,
after they have been secured together by fine-thread
sutures inserted close together. By this procedure
the natural shape of the mouth will be restored, as
is shown in Cases I. and II. The same method is
applicable where it is intended to extend the naturally-shaped mouth beyond its existing limits.

To avoid repetition, it may here be stated that the inhalation of ether was employed in all the operations reported in the subsequent pages.

The cases illustrating the subject of this volume are arranged under three classes.

First Class.—Loss of parts involving the face, and resulting from destructive disease or injury.

Second Class.—Congenital defects from excess or arrest of development.

Third Class.—Cicatricial contractions following burns.

FIRST CLASS.

LOSS OF PARTS INVOLVING THE FACE, AND RESULTING
FROM DESTRUCTIVE DISEASE OR INJURY.

CASE I.—*Reconstruction of the Mouth and Repair
of the Nose after the Loss of the Right Half of the
Upper Lip, the adjacent Portion of the Cheek and
Ala Nasi, together with the entire Right Superior
Maxillary Bone.*

CARLTON BURGAN, aged twenty, native of Maryland, and a private soldier in Company B, Permall
Legion, Maryland Volunteers. The following particulars of his antecedent history were furnished by
Robert F. Weir, M. D., Assistant Surgeon United
States Army, in charge of the Army General Hospital at Frederick, Maryland, where Burgan had been
a patient before coming to New York. He was
taken sick June 5, 1862, with rheumatic pains from
exposure to wet and cold while serving with his
regiment. He continued ailing till July 4th, when
he was sent to hospital and reported sick with
typhoid fever. On the 3d of August following he

was transferred to the General Hospital at Frederick.
About August 10th, although his general condition
appeared to be improving, a small black slough, attended with fœtor, showed itself upon the gum at
the root of the first upper bicuspid tooth on the
right side; the slough spread rapidly outward toward the cheek, and inward upon the roof of the
mouth. Both bicuspids and the canine tooth dropped
out. The outer surface of the cheek became swollen,
red, and glistening; the right eyelids swelled and
closed. The gangrene continued to spread until it
had destroyed the right half of the upper lip, the
adjacent portion of the cheek and ala nasi; it also
denuded the entire superior maxillary bone of the
same side. It was ascertained that, before coming
under Dr. Weir's charge, the patient had taken
within the space of two weeks, for the relief of ten-
derness of the right side, hydrarg. massæ, gr. lxv;
calomel, Ʒij; hydrarg. cum creta, Ʒj. During the
separation of the sloughs the fœtor was excessively
offensive. After they had all come away, healthy
action was established in the parts from which the
sloughs had separated, and patient's general health
steadily improved. On the 1st of October the right
superior maxillary bone having separated was removed entire, together with the vertical plate of the
os palati attached to it, and a narrow strip of bone
belonging to the left maxilla adjacent to the suture

in the median line where the two maxillæ are articulated with each other. (*See* Fig. 7.) The bone itself is deposited in the Army Medical Museum, at Washington, No. 557. By the middle of October the healing parts had so far contracted that the orifice of the cavity in the face had diminished to nearly one-half its original size upon the surface. December 31, 1862, Burgan was admitted into the New

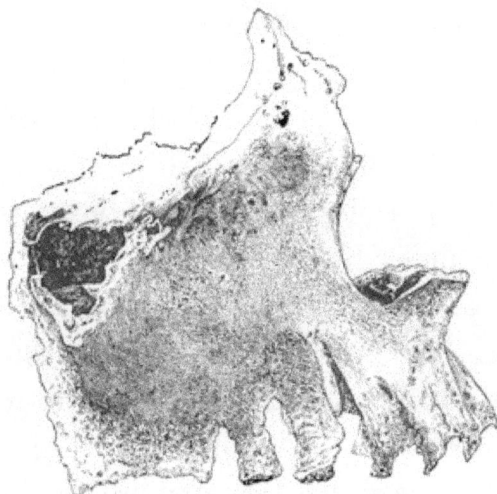

FIG. 7.

York Hospital, in a good state of health, with his face in the following condition: The right eye was sunken from atrophy of the eyeball, and the lids closed. The right half of the upper lip, the right ala nasi, and adjacent portion of the right cheek, together with the entire right superior maxillary bone, were gone, leaving an extensive opening into the

cavity of the mouth and the right nasal fossa. The
margin of the cavity at the surface was formed below
by the under lip, lengthened and extending obliquely
upward on the right side to the middle of the cheek,
where it terminated and adhered to the malar bone.
From this point the margin of the cavity continued
in a curved line below the inferior border of the
orbit to a point on the right side of the nose, one
finger's breadth below the inner canthus of the eye,
and thence downward to the apex of the nose along
the right side of the ridge. The columna nasi being
destroyed, the lower border of the left ala and the
rounded margin of the left half of the upper lip
bounded the opening in this direction. Only one-
third of the under lip was situated to the left of the
median line, while two-thirds were situated to the
right of it. The skin at the margin of the opening
overlapped it, and dipped somewhat into its cavity.
The inner wall of the cavity, toward the median
plane, was formed by the septum nasi, deflected
somewhat to the left side. The septum itself being deficient below from the absence of its cartilaginous portion, left the anterior extremity of the
left inferior turbinated bone and the lower orifice
of the nasal duct exposed to view. The roof of the
cavity was occupied by the inferior scrolled edge of
the middle turbinated bone belonging to the right
nasal fossa. The outer wall of the cavity presented

a smooth surface, continuous below with the inner
surface of the cheek. The tongue occupied the floor
of the cavity, and was exposed to view. The pala-
tine process of the os palati, which forms the posterior edge of the bony roof of the mouth and sup-
ports the velum, remained *in situ*, with its anterior

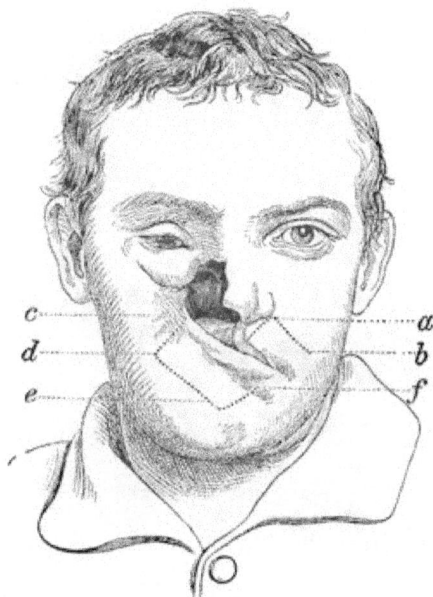

FIG. 8.

cicatrized edge stretching horizontally across the
middle of the cavity. The velum itself, thus supported *in situ*, performed its functions in deglutition.
The teeth belonging to the left upper maxilla, except
the middle incisor and one molar, were in place.
Articulation was very defective, and resembled that
of a person with a bad cleft palate. (*See* Fig. 8.)

In devising a plan for the repair of this extensive loss of parts, it was judged indispensable, as a prerequisite step to any surgical operation, that some artificial substitute should be adapted to the cavity of the mouth that would supply the place of the lost maxillary bone, and afford a solid support to the soft parts that would have to be transposed for the reconstruction of the mouth and the closure of the cheek and nostrils. Mr. Thomas B. Gunning, a skill-

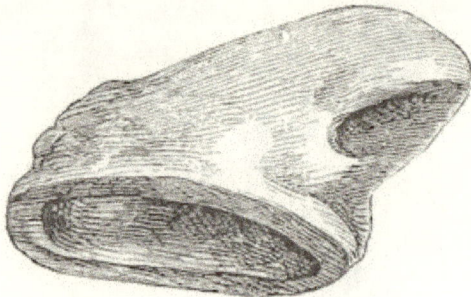

Fig. 9.

ful dentist of this city, to whom the case was submitted, generously undertook the execution of this delicate and difficult piece of work. The fixtures which he adapted were made of vulcanite, and consisted of two principal pieces, superposed, when in place, one above the other. The upper, or nose-piece, occupied the nasal fossa, and filled out the right half of the nose. It was hollow, and open in front and behind for the free passage of air. (Fig. 9.) The lower, or palate-piece, occupied the roof of the mouth, and supported upon its upper surface the nose-piece. It consisted of a plate stretching across the roof of the mouth and supplying the dental arch at its margin on the right side, together with the teeth belonging to it. Its left margin took support from the existing teeth of that side, some of which it embraced. (Fig. 10.) The surfaces of both pieces, where they came in contact with the walls of the nasal and buccal cavities, were channeled with fur-

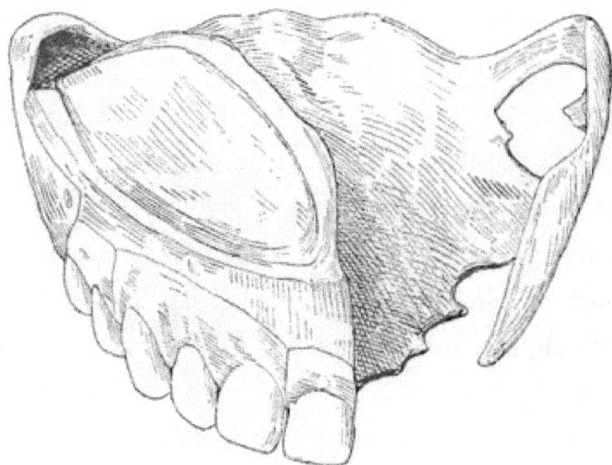

Fig. 10.

rowed lines to facilitate the flow of the secretions back into the fauces. Their accurate adjustment to each other, and to the cavities they were adapted to occupy, permitted them to be worn without causing any irritation. The light and indestructible nature of the material of which the pieces were constructed adapted them admirably to their present use. On trial patient found he could wear them constantly with comfort, and could remove and replace them at pleasure. The improvement of his articulation and the increased facility of mastication they effected when in place afforded him the highest satisfaction.

The requisite preparations being now completed, it was decided to perform the first operation on the 26th of March, 1862. In order to guard by every possible precaution against erysipelas, and other preventable complications, the patient was placed in an outbuilding on the hospital premises which had not been in use for several weeks, and where he and his attendants would be the sole occupants.

First Operation.—Was performed as follows. The First Step of the operation was to prepare the left half of the upper lip. The lip being held on the stretch, the buccal mucous membrane was incised along the line of its reflection from the upper jaw to the lip and cheek as far outward as the molar teeth. The lip itself was then divided through its entire thickness from the point where it joined the left ala nasi, on a line parallel with the lip border, outward to the middle of the cheek. The lip flap thus formed was trimmed square at its free extremity. The Second Step was to prepare the redundant under lip, so as to employ it for supplying the deficient right half of the upper lip. This was done

according to the method described on page 26. After
the newly-adjusted parts had been secured by the
requisite sutures, the reconstructed right half of the
mouth assumed a circular shape, and stood prominently forward. The Third Step had for its object
to close the open space in the right cheek, resulting
from the transposition of the parts effected by the
Second Step of the operation. It was executed as
follows: A transverse incision was carried through
the entire thickness of the right cheek on a line with
the commissure of the mouth, as far outward as the
anterior edge of the masseter muscle, and beyond it
only through the skin covering the muscle. In order
to glide forward the divided cheek, the buccal mucous
membrane alone was divided along the anterior edge
of the masseter, above and below the transverse incision. This allowed the cheek to come forward and
meet the outer margin of the transposed under-lip
flap. Their confronting edges, after being pared and
matched to each other, were secured in exact coaptation by silver-wire and fine-thread sutures, inserted
close together. Before closing the parts, the numerous ligatures that had been applied to bleeding vessels were loosely twisted into a skein, and brought
out to the surface at the outer angle of the transverse
incision. The operation now completed had occupied
nearly three hours, owing in part to the frequent interruptions necessary to maintain the effect of the

ether. No adhesive plaster was applied. Warm-water dressings were directed to be kept to the parts.

March 23*d.*—Progress favorable; pulse 90. Inflammatory tumefaction moderate. The yarn upon the pins in the upper lip being sunken in the swollen parts, it was removed, and, after allowing the underlying constricted surface to recover itself, fresh yarn was applied. Liquid nourishment and drinks were taken without difficulty. Patient experiences no uneasiness from the presence of the artificial fixtures. Injections were employed to cleanse the spaces between and around them.

28*th.*—General condition favorable; pulse 90. Swelling extended to the right side of the neck, below the jaw. No indications of deficient vitality existed at any point. Removed the alternate thread sutures and the silver-wire sutures from the lip.

29*th.*—Progress still favorable; pulse 104. Copious suppuration is taking place from the track of the ligatures, and is somewhat offensive. Removed additional thread sutures. At the junction of the two halves of the upper lip the new adhesions were in danger of giving way, owing to tension. To prevent it a fresh pin was inserted between the two old ones, which were then removed.

30*th.*—Several ligatures have come away. Suppuration diminishing and no longer offensive; removed the sutures from the vermilion border.

31st.—Removed the supplementary pin suture, and supported the upper lip with adhesive plasters.

April 9th.—Parts had all healed, and dressings were no longer required. The result is shown by Fig. 11.

Fig. 11.

The parts involved in the preceding operation having regained their natural suppleness, a Second Operation was performed April 23d, for the purpose of restoring to the right half of the mouth its natural angular shape, and lengthening it at the same time.

Second Operation.—Was performed according to the method described on page 28. At the expiration of twenty-four hours after the operation the alternate

thread sutures were removed, and on each succeeding
day others were got rid of, as fast as they could be
dispensed with, until the fourth day, when healing
was complete, and the last suture was removed. The
result was satisfactory, and is shown by Fig. 12.

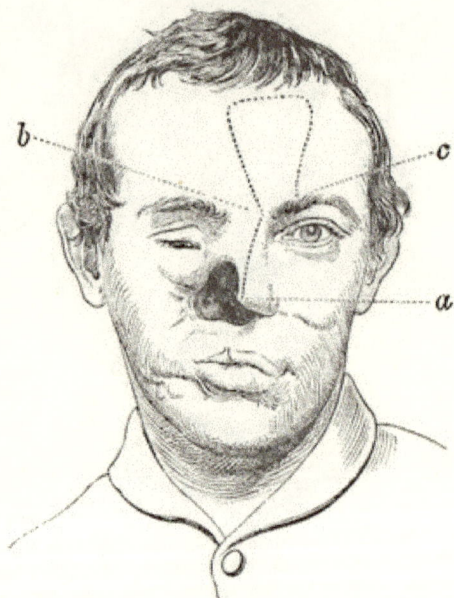

Fig. 12.

The dental fixtures having been worn constantly
for more than six weeks, it was found necessary to
remove them, in order to relieve the natural teeth
of the left upper maxilla from the too great pressure
to which they had been subjected by the insertion
of pieces of sponge between the right cheek and the
artificial teeth belonging to the palate-plate; these
sponges had been employed to resist the contraction

of the newly-cicatrizing parts. During the succeeding eight weeks, in which the dental fixtures were
left out of use, the parts underwent changes which
require some particular notice. For instance, wherever the incisions involving the right cheek had divided the lining mucous membrane, tense salient
bands had formed. One of these crossed the inside
of the cheek horizontally, another extended upward
and backward, deep below the orbit; both bands
were continuous in front with the upper border of
the transposed under-lip flap, which now constituted
the right half of the upper lip. This same border
also formed the lower margin of the still-existing
opening into the right cheek, which corresponded to
the deficient right ala nasi. In attempting to open
the mouth, these bands were put tightly on the
stretch, and, proceeding from a point above the outer
incisor tooth of the left upper maxilla, they prevented a separation of the jaws beyond a space sufficient to admit a finger edgewise between the teeth.
A firm adhesion between the cheek and lower jaw,
on the right side of the chin, was an additional obstacle to the separation of the jaws. This condition
of the parts was an insuperable obstacle to the re-
introduction of the dental fixtures, and required to
be overcome preparatory to any further operation.
This was done without anæsthesia in the following
manner: A pair of blunt-pointed scissors, curved

flatwise, was introduced into the mouth, guided by
the forefinger of the left hand, and with them the
upper band was divided at its remotest point under
the orbit. The horizontal band was also freely divided at its farthest point, and the adhesions between
the cheek and lower jaw were also freely liberated.
Considerable relaxation of the parts was thus obtained, but not sufficient for our purpose, until the
left half of the upper lip was a second time divided
through its entire thickness horizontally along the
line of the cicatrice left by the first operation. This
had the desired effect, and the replacement of the
dental fixtures was now accomplished by Mr. Gunning himself, who was present on the occasion.

Third Operation.—Performed June 18th, at 3 P. M.
The object of it was to close the remaining opening
in the right cheek, and cover the adjacent side of the
nose with a patch of skin, to be taken from the left
half of the forehead. The opening, as shown by
Fig. 12, involved the right half of the nose below
the os nasi, and the adjacent cheek as far out as
its middle. The lower border of the opening extended in a line from the inferior edge of the left
ala nasi horizontally across to the middle of the
right cheek, and formed also the upper margin of the
newly-transposed under-lip patch. In order better
to adjust the edges of the patch of skin that was to
be brought down from the forehead, the skin at the

margin of the opening was dissected up from its
underlying connections sufficiently to permit it to be
everted.   At the outer margin of the opening, which
corresponded to the anterior 'edge of the masseter
muscle, the skin, which dipped somewhat into the
cavity, was detached more extensively, so as to permit it to be glided forward, and thus contribute to
the closure of the opening.   A pattern of the shape
of the opening, but somewhat exceeding it in size,
as an allowance for shrinkage, was then cut from
oiled silk and laid upon the left half of the forehead,
in an inverted position, with its broadest part directed upward, and its narrowest applied above the
inner half of the left eyebrow.   An incision was then
commenced at the tip of the nose, Fig. 12, *a*, and carried upward along its dorsum, skirting the margin
of the opening, and extending beyond it obliquely to
the inner extremity of the left eyebrow, *b*, where it
encountered the right margin of the pattern, and was
continued thence around its entire circumference till
it reached the middle of the left eyebrow, *c*.   The
included patch of skin was then dissected up from
the pericranium, and left attached below at the supraorbital ridge by its pedicle, which measured one inch
in breadth.   In order that the patch from the forehead should, after its transfer, have its under surface
lie in immediate contact with raw surface, the skin
intervening between the eyebrows was dissected up

and removed. The patch itself was then brought round edgewise from left to right, and above downward, till it reached its new location on the right side of the nose and cheek, where it was adjusted with its straight edge along the entire length of the dorsum nasi, and secured there by three pin sutures, inserted at equal distances apart, and additional intermediate fine-thread sutures. What had now become the right border of the patch in its new locality, was adjusted to that part of the margin of the opening which involved the cheek, and was secured to it by fine-thread sutures, inserted close together. The inferior and outer angle of the patch, which, after its junction with the cheek, would bound the nasal orifice on the right side, was adjusted as follows: At the angle where the outer and inner borders of the opening in the cheek met, a strip, half an inch long, was raised from the inferior border, and left attached toward the median line. The angle of the patch was then adjusted in its place, and the detached strip applied to its inner raw surface, and made to line it. The edges of the transverse incision, by which the left half of the upper lip had been liberated the second time, were again secured to each other by sutures. The ligatures that had been applied to bleeding vessels were brought to the surface at the nearest point of exit. The raw surface left upon the forehead was coated with a collodion crust, in the manner described on page 13. Lint was stuffed between the cheek and jaw, on the right side of the chin, to prevent adhesion from again taking place. Warm-water dressings were applied to the forehead and cheek, and dry warmth to the nose, by means of a bat of cotton and a vial of hot water suspended in contact with it. Although the operation occupied nearly three hours, it was not followed by any unpleasant effects from the long-continued administration of ether.

June 19*th.*—Patient passed a comfortable night, and obtained some sleep. The transferred patch was cool to the touch, the difference of temperature being 10°. Covered the surface with threads moistened with spirits of turpentine, and continued the same applications.

20*th.*—Progress favorable; temperature of the patch was restored and well sustained. The confronted edges of the patch and adjacent skin were alike tumefied, and showed no difference in their degree of vitality. Changed the yarn on the pins.

22*d.*—Patient continued to do well. Removed the only remaining pin from the tip of the nose, and the alternate thread sutures at other points.

23*d.*—Removed other sutures; all the ligatures had come away.

24*th.*—The edges of the patch had united at all points by first intention.

4

26th.—The last suture was removed. Strips of adhesive plaster were applied to support the newly-healed parts.

27th.—This being the ninth day, the collodion crust separated and dropped off, exposing a healthy granulating surface, nearly on a level with the surrounding skin. Scarcely any inflammatory redness or swelling had at any time appeared around the margin of the crust.

July 24*th.*—Continued favorable progress. The transferred patch in its new connections had undergone a process of hypertrophy, and obtained a uniform thickness of half an inch. The adjacent skin, to which it was united, participated in the same thickened condition to a distance of one-quarter of an inch from the line of junction. The surface of the patch was paler than the neighboring skin; the sore upon the forehead had progressively diminished in size. Upon the upper part of the nose, and between the eyebrows, the skin was gathered into a bulging fold, in consequence of the doubling of the pedicle of the patch upon itself in its transfer from the forehead. The conspicuous disfigurement thereby produced was now to be removed by a

Fourth Operation.—Performed August 8th. Two parallel incisions, commencing one on either side of the dorsum nasi, were continued upward on the forehead, where they converged and met so as to include the bulging fold of skin in a tongue-shaped strip, which was itself dissected up from its under-lying connections. Narrow strips of skin were re-moved from the edges of the wound on either side, to make room for the strip, which was spread out and replaced in an advanced position, and secured by numerous sutures to the adjacent skin on all sides. With a view to reduce the excessive thickness of the new ala nasi, at its inferior margin, where it bounded the orifice of the nostril, a prism-shaped strip, pene-trating deep between the outer and inner surfaces, was excised, and the wound closed by fine-thread sutures. Both these operations were followed by primary union, and did well.

September 1*st*.—The wound surface upon the fore-head had healed, leaving a cicatrice scarcely exceed-ing one-third the size of the original wound. The shrinkage which the nose-patch had undergone had produced a deep furrow along the cicatricial line, occupying the lower third of the ridge of the nose. The furrow terminated below, at the apex of the nose, in a deep, unsightly notch. To remedy this disfigure-ment was the design of a fifth, and final, operation.

Fifth Operation.—Performed October 27th, as fol-lows: Two parallel incisions were made, one on either side of the furrow, penetrating deep in converging planes, so as to include both sides of the furrow and the notch below. The opposite edges of the wound

were secured in accurate contact by two pin sutures
and three fine-thread sutures. Within six days the
sutures were all removed and perfect union obtained.
A great improvement in the appearance of the nose
was the result of this operation. In June, 1864, Burgan was in the enjoyment of good health, and had
for several months preceding discharged the duties
of an assistant-nurse in a large ward of the hospital.
The hypertrophied condition of the nasal patch still
persisted giving to the right half of the nose a
plump form. When the surface was pricked, the
sensation was no longer referred to the forehead as
at first, but to the actual seat of irritation. The cicatricial bands on the inside of the right cheek had
been kept from again contracting by the persevering
efforts of the patient, who had faithfully executed
the directions given him, to introduce one or more
fingers into the mouth and stretch the bands to their
utmost limit, and to repeat the operation several
times daily. The artificial substitute for the right
maxilla, which consists of a palate-plate that covers
the roof of the mouth, and supplies the lost teeth of
the right side, is the only piece now worn, the nasal
piece having been dispensed with for a long time.
It is worn constantly with entire comfort, and can
be removed and replaced at pleasure. With it patient is able to masticate all kinds of food, and articulation, which without it was scarcely intelligible,

now betrays very little defect. In May of 1871 the
author visited Burgan at his home, in the suburbs
of Baltimore, and found him a married man, and
father of two children. He enjoyed good health,
and was pursuing a laborious occupation. He still
wore constantly the same palate-plate that had been

Fig. 13.

adapted in March, 1863. Time had considerably
improved the condition of the right cheek and nose.
The cicatricial lines on the surface of the cheek were
no longer elevated ridges, but had shrunk into shallow furrows, and become much less conspicuous.
The cheek itself was pliable at all points, and the
right half of the nose retained its plump shape, al-

though it had lost its excessive thickness.   Fig. 13
is from a photograph, taken at Balch's Studio in Baltimore, in 1871.

CASE II.—*Reconstruction of the Mouth and Repair
of the Nose after the Loss of the Right Half of the
Upper Lip, the Adjacent Portion of the Cheek, and
Right Ala Nasi.*

JOHN MICHÆLIS, aged eleven, of German parentage, a resident of Jamaica, L. I., was admitted into
St. Luke's Hospital in April, 1866. The loss of parts
sustained in this case happened in the progress of
malarial fever, with which he was attacked in the
month of September preceding. After a fortnight's
treatment, during which, according to his mother's
statement, he took a great deal of medicine, sloughing invaded the right half of the upper lip and adjacent cheek. At the expiration of about ten weeks
the parts from which the sloughs had separated were
healed, and had assumed the condition in which they
were at the time of his admission to the hospital.
It was as follows: The right half of the upper lip,
the adjacent portion of the right cheek, and the right
ala nasi, were gone. The upper front teeth and
gum-surface belonging to the right half of the upper
maxillary bone were bare and exposed to view. A

supernumerary canine tooth emerged from the gum above the level of the other teeth, in the space between the outer incisor and canine teeth, and formed a conspicuous feature in the disfigurement of the face. The gum-surface above these teeth was incrusted with a brownish scab, from underneath which healthy pus escaped on pressure. On the right side of the nose a semicircular notch occupied the place of the ala that had been destroyed. The upper edge of the notch corresponded to the lower margin of the os nasi. The septum and columna nasi were entire. The cicatrized margin bounding the deficiency in the cheek was sunk, and closely adherent to the upper maxilla above the teeth. The lower-lip border was lengthened on the right side, and stretched obliquely outward and upward to the alveolar border of the upper jaw, where it adhered to the gum-surface above the bicuspid teeth. From this point of adhesion a linear cicatrix one inch long extended outward, and a little downward, half across the cheek, depressing the surface and adhering to the underlying parts. The median line divided the under-lip border unequally, three-fifths of the border lying to the right and two-fifths to the left of it. The remaining left half of the upper lip was shrunk, at the point where its vermilion border terminated below the columna nasi. The motions of the lower jaw were unrestricted. The patient's general health

was good. The outer incisor, canine, and supernumerary canine teeth of the upper jaw were extracted preparatory to an operation. Fig. 14 shows
the condition of the face just described.   .

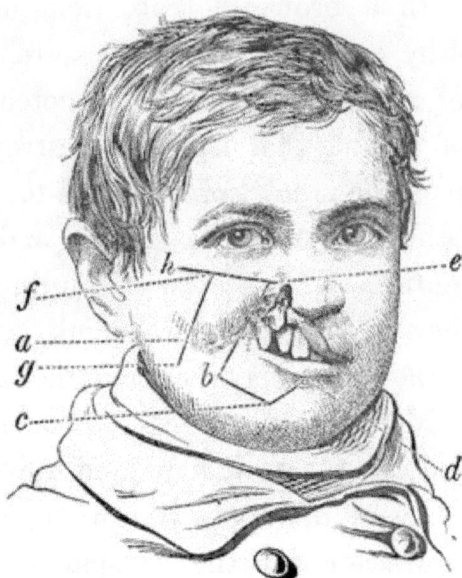

FIG. 14.

First Operation was performed May 18th as follows: The left half of the upper lip, being held upon
the stretch, was detached from the jaw by an incision of the buccal mucous membrane, carried along
the line of its reflection from the jaw to the lip and
cheek, and extended outward as far as the molar
teeth, and also upward on the level of the periosteum
toward the orbit. This permitted the lip and cheek

to be glided over toward the right side. A strip of the vermilion border, one inch in length, was pared away from the extremity of the half lip and left attached temporarily to it. Material with which to supply the deficient half of the upper lip was next obtained from the redundant right half of the under lip. It was done according to the method described on page 26, and also employed in Case I. (*see* page 38). The quadrilateral flap (*a, b, c*) thus formed from the under lip was adjusted by its free extremity, after being brought round edgewise to the left half of the upper lip, and the two were secured together in a vertical line below the columna nasi by two pin sutures and several fine-thread sutures. The open space in the right cheek, remaining after the transposition of the parts as just described, was now closed in the following manner: An incision, one inch and a half long (*e* to *f*), was carried transversely across the right cheek, on a level with the middle of the nose, to a point one inch below the outer canthus of the eye, and thence downward in a curved line (*f* to *g*), with its convexity directed backward to a point within one inch of the margin of the lower jaw. The quadrilateral flap of skin (*b, e, f, g*), lined with mucous membrane, was dissected up from the jaw, but left attached below. It was then slid forward edgewise till its anterior border met the outer border of the newly-transposed under-lip flap, and the two were matched together and secured by pin sutures and fine-thread sutures, distributed so as to effect the most exact adjustment of the edges to each other. In order to cover the bare surface remaining on the cheek after the transfer just made, the incision e, f, crossing the cheek, was prolonged to the point h, upon the temple, and the angle of skin, h, f, g, included between it and the vertical incision, f, g, was dissected up from its underlying connections sufficiently to allow it to be drawn forward and cover the bare surface, and there be adjusted with sutures. The reconstructed right half of the mouth had now assumed a circular and pouting form, a defect which it was intended to remedy by a subsequent operation. The same after-treatment was employed in this case as in the preceding one, and with a like good result. There was no lack of vitality, nor sloughing, at any point. The inflammatory tumefaction was moderate, and the sutures were removed in succession as they could be dispensed with. On the tenth day all had been got rid of, and patient was out of bed and able to go about. The result of the operation is shown in Fig. 15. The notch involving the right ala of the nose, constituting as it did a part of the original defect, now formed with the adjacent cheek and upper lip a foramen capable of admitting the end of a finger. The closure of this foramen was next to be

attempted by a second operation, which was performed on the 18th of June, as follows:

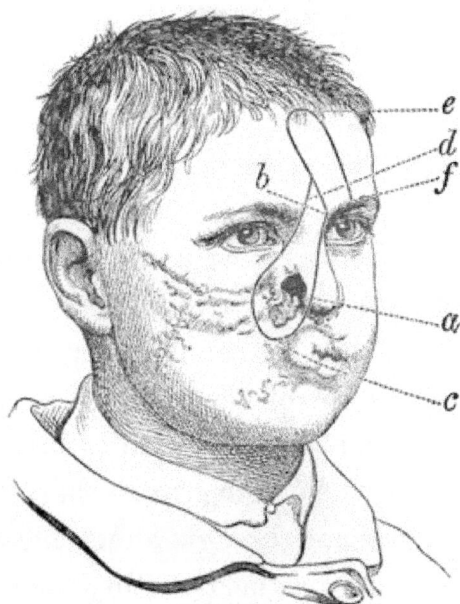

Fig. 15.

Second Operation. — The edge of the foramen,
where it involved the cheek and upper lip, was
pared and everted. On the right side of the apex
nasi an incision, skirting the edge of the notch, was
carried up obliquely over the dorsum of the nose to
the inner extremity of the left eyebrow, *a* to *b*. A
second incision, commencing at the opposite edge of
the foramen upon the right cheek, was carried upward upon the right side of the nose parallel with
the first incision to the inner extremity of the right
eyebrow. The skin included between these two in-

cisions was dissected up from the side of the nose, and between the eyebrows, and removed.' A pattern of the shape of the bare surface thus prepared was cut from oiled-silk, and laid upon the forehead above the inner half of the left eyebrow. The incision previously made, that terminated at the inner extremity of the left eyebrow, was then continued upward along the right edge of the pattern, and onward around the rest of its margin to the middle of the eyebrow. The included patch of skin was then dissected up from the pericranium, but left attached below at the margin of the orbit where its pedicle derived a vascular support from a branch of the ophthalmic artery emerging from the orbit. The patch of skin was now brought down edgewise from left to right to its destined locality, and its edges accurately adjusted to the edges of the space prepared for it, by pin sutures and fine-thread sutures inserted close to each other. A short piece of gum-catheter was inserted in the nasal orifice, and maintained there for the purpose of establishing a permanent aperture. The opposite edges of the bare surface upon the forehead were approximated, and held in contact by sutures and strips of adhesive plaster.

' This portion of skin it would have been better to save and utilize for covering the bare surface on the forehead, as was done with great advantage in subsequent similar operations.

The after-treatment was the same as already described in previous operations. The subsequent progress was favorable. The transposed patch maintained its vitality, and united by primary adhesion
at all points except at its lower extremity, where,
after suppurating for several weeks, it healed at

Fig. 16.

length by the granulating process. The wound upon
the forehead failed to heal by adhesion, but finally
healed by granulation, the edges being held approximated during the process by adhesive-plaster, so that
the resulting cicatrice was reduced to almost linear
dimensions. The result of the operation is shown
by Fig. 16.

On the 31st of July, patient was allowed to go to his home to recruit. After his return to the hospital much improved in health, a third operation was undertaken for the purpose of restoring the newly-reconstructed right half of the mouth, which had assumed a circular, protruding form, to its natural angular shape, and at the same time to lengthen the mouth on the same side.

Third Operation was performed September 26th, according to the method described on page 28. An operation was also performed at the same time for the removal of a conspicuous ridge of skin occupying the upper part of the nose, and the space above between the eyebrows. It had resulted from the folding upon itself of the pedicle of the patch of skin that had been brought down from the forehead. Two curved incisions were carried, one on either side of the elongated ridge, and made to meet at its opposite extremities. The elevated skin included between the incisions was removed, and the opposite edges of the wound were brought together and secured by sutures. Primary adhesion followed in both operations, and an excellent result was obtained.

A great improvement of the external appearance of the right half of the face had been effected by the operations already performed, but there still remained a serious defect dependent on cicatricial contractions on the inside of the right cheek, which held the jaws together and prevented the teeth from being separated in front more than half an inch. A single cicatricial band of mucous membrane on the inside of the right cheek, close to the angle of the mouth, presented the greatest resistance to the separation of the jaws. With the view of remedying this defect it was proposed to lengthen the mouth, which was scanty on this side, by extending the angle farther toward the cheek. In doing this the constricting cicatricial band on the inside of the cheek would be divided, and the jaws thereby liberated. For this purpose a fourth, and final operation, was performed in July, 1868.

Fourth Operation was a repetition of the preceding third operation, and secured the desired result, so that the mouth became more symmetrical in shape, and the teeth could be separated one inch in front.

Before leaving the hospital the patient was instructed to insert wedges of wood between the upper and lower molar teeth, and wear them as long at a time, and as often daily, as possible, with the view of increasing the power of separating the jaws. A progressive improvement was afterward observable. In April, 1871, the mouth could be opened and the jaws separated to the fullest extent. The right cheek was pliable, the cicatricial lines on its outer surface

FIG. 17.

had shrunk and become much less conspicuous. Fig. 17, copied from a photograph taken at this time, shows the final result.

CASE III.—*Reconstruction of the Nose and Mouth after the Loss of the Nose and entire Upper Lip.*

JANE TUCKER, aged twenty-six, native of Ireland, unmarried, admitted into St. Luke's Hospital, March 10, 1866. The antecedent history of her case is as follows: When about seven years old, sores formed on the inside of her nose, from a practice she was addicted to of constantly picking her nose with her fingers. Destructive ulceration followed, which extended to the upper lip, and finally resulted in the condition in which she was when admitted into the hospital. It was as follows: The nose below the nasal bones was sunk to the level of the cheeks, and both nostrils were blocked up, the only external opening being a small perforation situated in the median line on a level with the floor of the nasal cavity, and of a size capable of admitting an ordinary probe. The ossa nasi still remained *in situ*. The upper lip was entirely gone, leaving uncovered the upper front teeth and adjacent gum-surface within an angular space bounded by two lines which diverged from a central point on a level with the floor of the nostrils, and descended to the angles of the mouth. These lines also marked the limits of the sound skin on either side. Both angles of the mouth were entire, and the under lip was of ample dimensions, being even somewhat pendulous. Patient's general health was good. Fig. 18 shows the condition just described.

First Operation.—The design of the operation was to reconstruct the upper lip from the redundant material of the lower lip. It was performed on the 28th of March, as follows: An incision, commencing at the median line, on a level with the floor of the nasal cavity, was carried outward and downward on

5

both sides of the face in a curved line so as to circumscribe both angles of the mouth, and terminate
at a point below the junction of the middle and
outer third of the under lip (*a* to *b*, *a* to *c*, Fig. 18).
These incisions divided the entire thickness of the
cheeks and · lip, and in their course were kept at a

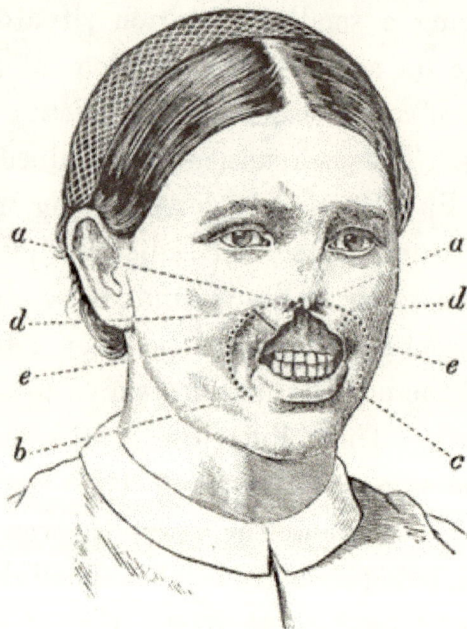

Fig. 18.

uniform distance of one inch and a quarter from the
angles of the mouth and the under-lip border. The
flaps thus formed on either side were brought toward
each other edgewise, and their ends, after being
pared and made straight, were adjusted to each other
on a vertical line in the median plane, and secured

by three pin sutures and intermediate thread sutures.
What had now become the upper border of these
united flaps was adjusted to the opposite cut edge
of the skin above, and secured by sutures.   In order
to close the open spaces between the newly-transposed flaps and the cheeks on either side, from which
they had been detached, it was necessary to liberate
the cheeks by incising the mucous membrane on the
inside of the mouth, above and below, along the
line of junction between the cheeks and the jaw-
bones.   This permitted the cheeks to be brought
forward to meet the flaps encircling the new mouth,
and close up the spaces between them.   This adjustment was secured by pin sutures and thread sutures
distributed so as to afford support to the parts at
all points, and effect the most exact coaptation of
their edges.   No bare surface was left uncovered.
The reconstructed mouth necessarily assumed a circular and pouting shape.   Primary union followed
at all points, except where the flaps came together
in the median line below the nasal cavity.   Here it
failed in consequence of sloughing of the newly-
united edges, which separated from each other except
at the upper third of their line of junction.   This
separation, including as it did the lip border, left the
upper front teeth again uncovered as before, though
not to so great an extent.

Patient remained in the hospital till the 24th of

July, when she was discharged in ordinary good
health. On the 13th of March, 1867, she was admitted into the New York Hospital to undergo further operations. It was now proposed to remove the
obstruction of the nostrils, and afford a free passage
for air through them, thereby enabling the patient
to have an artificial nose adapted.

Second Operation.—A vertical incision was carried from a point midway between the eyebrows
downward upon the nose to a point on a level with
the floor of the nasal cavity; a transverse incision
crossed the lower extremity of this vertical incision,
and extended one inch on either side of it, the two
forming together an inverted letter-T-shaped incision.
The angular flaps of skin on either side were dissected up, and the sunken parts blocking up the
nasal orifice were cleared away from between the
ascending nasal processes of the superior maxillary
bones, which form on either side the lateral bound-
ary of the nasal fossæ. The skin was then pared to
correspond to the bony margin of the new opening
and left to cicatrize, which it did in due time, leaving a permanent aperture with rounded edges, and
of the shape of an inverted heart, measuring three-
quarters of an inch in its vertical, and five-eighths
of an inch in its transverse, diameter. (*See* Fig.
19.)

Third Operation was performed on the 14th of

April, 1867. Its object was to attempt once more the reconstruction of the upper lip. Although the first operation on the mouth had been only partially successful, an important advantage had been gained by securing a covering of skin between the upper front teeth and the nasal orifice. The steps of the operation were the following: An incision was commenced about half a finger's breadth below the nasal orifice and carried along the margin of the newly-transposed skin, and continued on both sides outward and downward in a curved direction, so as to circumscribe the angles of the mouth, and terminate at corresponding points on each side of the chin, below the junction of the middle and outer third of the under-lip border. These incisions were kept at a distance of one inch from the angles of the mouth and the under-lip border, and divided the entire thickness of the cheek. The two lateral flaps thus formed were brought together edgewise and made to meet in the median line, where their ends, after being squared, were adjusted together by sutures. The outer (i. e., toward the cheeks) edges of these united flaps were readjusted to the opposite edges of the cheeks, from which they had just been detached, by the same procedure as had been employed in the first operation, of which this was almost an exact repetition. As a result of this operation the mouth again assumed a circular and pouting shape.

The subsequent management of the sutures was the same as has been already described in previous operations. The edges of the flaps at their junction in the median line below the nasal aperture, and at their junction with the cheek, to the distance of one inch on either side of the median line, assumed a livid, ashy appearance, and failed to unite by primary adhesion.

Erysipelas of a mild form showed itself on the 16th of April, attended with moderate swelling; on the 19th it had disappeared. The sloughy edges of the wound having cleared off and begun to granulate, it now became imperative to maintain them in contact during the process of union by granulation. This was effected by sutures, inserted at every point where they could afford support. General invigorating treatment was also enforced, such as sulph. quiniæ, iron, wine, and generous diet.

From this date onward to the 7th of May, the most assiduous care was required to support the edges in contact with each other. Sutures were renewed at every available point once in two or three days, and removed before they had time to produce suppuration. Strips of adhesive-plaster were also employed for support. Union was at length secured at all points. The left half of the new upper lip, having suffered most from sloughing, is somewhat shrunk, and also slightly notched, at its border, where it joins the right half in the median line. The result thus achieved, though not perfect, has nevertheless brought the parts into a condition in

Fig. 19.

which it will be possible by another operation to restore the mouth to its natural shape.

Fourth Operation.—The mouth having assumed, as a result of the preceding operation, a circular form, it was the design of the fourth operation to restore it to its natural angular shape. It was executed in July, 1867, on both sides of the mouth simultaneously, according to the method described on page 28, and was followed by a satisfactory result, as shown by Fig. 19.   After leaving the hospital in January, 1868, the patient had an artificial nose, made of vulcanite, adapted to her face.   She wore it for some time, but without any satisfaction, and finally cast it aside in disgust.

On the 8th of January, 1869, Jane was again admitted into the New York Hospital, to undergo an operation for the construction of a nose, the material for which was to be taken from the forehead.

Fifth Operation.—As a preliminary step to this operation, a nose was modeled in wax by a sculptor upon a plaster cast of the patient's face previously taken for the purpose.   This served as a pattern from which to determine the size and shape of the patch of skin that would be required.   The external nasal aperture was then prepared by carrying an incision along its margin on both sides, and dissecting up the outer edges of these incisions toward the cheeks, so as to adapt them for coaptation with the lateral edges of the patch of skin that was to form the new nose.   A pattern, cut from oiled-silk and shaped upon the model of the nose, was laid upon the forehead in an inverted position, with its pedicle of attachment applied above the left supraorbitar notch, and its long axis lying obliquely upon the forehead and inclining to the right of the median line.   An incision, involving the thickness of the skin, was then carried around the margin of the pattern, and the included patch of skin, after being dissected up from the pericranium, was left attached below at the margin of the orbit. The skin covering the space between the eyebrows and the ossa nasi was displaced, to afford a continuous raw surface to which to apply the under surface of the patch, when brought down from the forehead. This portion of skin, however, was left attached at the inner extremity of the right eyebrow and reserved for subsequent use. The patch itself was then brought down edgewise from the forehead to its destined location over the nasal aperture, to the edges of which (prepared as described above) the edges of the patch were accurately adjusted and secured by numerous sutures. The bare surface left upon the forehead was covered at its lowest part by the reserved portion of skin that had been displaced from between the eyebrows. The upper portion of the bare surface was treated by covering it with a collodion crust (*see* page 13). The cavity of the newly-constructed nose was stuffed with lint to maintain it plump and in good shape. The patch, in its new locality, underwent no apparent change in color or temperature. Primary union followed at all points except on the right side of the nose, high up, where suppuration took place between the edges of the wound, for a distance of three-quarters of an inch.

Several weeks after the operation just described,

an attempt was made to close the opening by paring its edges and securing them in contact by sutures. To my great disappointment sloughing followed, and an enlargement of the size of the opening was the consequence. After a further delay of several months the opening still existed, and measured three-fourths of an inch in its vertical, and half an inch in its transverse, diameter. The closure of this opening was next attempted by a

Sixth Operation.—It was executed as follows: The edges of the opening were pared and everted. A patch of skin, of suitable shape and size, was raised from the forehead above the opening, and left attached below at the supraorbital notch on the right side. It was then turned down and adapted to the opening, with its cuticular surface directed toward the nasal cavity and its raw surface outward. Fine-thread sutures were inserted in close proximity to each other, to secure the most accurate adjustment of the edges. In making this transfer it was necessary to stretch the flap across an intervening space of sound skin of about a finger's breadth.

At the end of three weeks, the patch, which had become ingrafted in its new locality, was divided at the upper margin of the opening and its pedicle turned up and replaced in its original site, there to contribute toward covering the bare surface upon the forehead from which it had been taken. A complete closure of the opening, however, was not effected by this operation, as a small aperture still remained at the upper margin, allowing the passage of air through it. The outer raw surface of the transplanted patch cicatrized, and adapted itself so exactly in its new locality that its outline could scarcely be distinguished.

In the month of June, 1870, Jane again became a patient at St. Luke's Hospital, and two successive operations were performed for the purpose of closing the opening still existing on the right side of the nose. The first of these operations consisted in paring the edges of the opening, detaching them extensively from the underlying parts and securing them in contact by sutures. This failed from defect in the adhesive process.

After waiting a sufficient length of time, and allowing the edges to cicatrize, a second attempt at closure was made as follows: The edges of the opening were seared with the actual cautery, and, as soon as they had assumed a granulating condition, a slender tenotomy-knife was inserted under the skin, at a distance of three-fourths of an inch from the opening, and swept around flatwise in every direction so as to liberate the skin from its underlying connections. In order to maintain the opposite granulating edges of the opening in contact, and thus facilitate their adhesion, a beaded silver-wire suture (*see* page 17) was inserted crosswise. This expedient, although it secured the desired contact of the edges, failed to effect a complete closure of the opening. What remained of it, however, ultimately contracted down to so small a size that the patient was contented to wear a small plug of wax which kept it closed, and was scarcely noticeable. She also wore habitually a plug of lint in the nasal cavity to keep it distended and maintain the nose more prominent.

FIG. 20.

In the summer of 1871 the patient's condition was ascertained to be as follows: The mouth retained its natural shape and dimensions, and the lips performed all their functions. The nose had shrunk considerably since it was first reconstructed, though for a year past it had remained stationary. The voice was nasal in a very marked degree. The nose, though defective in shape, was much less repulsive to the eye than an artificial nose would be. Fig. 20, showing the final condition of the face, is from a photograph, as are the other two.

The patient was again examined June 23, 1875. The condition of her face remained much the same as described above, without having undergone any perceptible deterioration.

CASE IV.—*Reconstruction of the Mouth after the Loss of the entire Under Lip, and a portion of the Inferior Maxillary Bone.*

HUGH B., aged thirteen, a resident of the city, of rather slender constitution, though ordinarily enjoying good health, came under surgical treatment in October, 1869, for the relief of the condition of his mouth, of which his mother gave the following account: When six years old he was attacked with scarlet fever, and became dangerously ill. The eruption was scanty, and in other respects the disease was irregular in its development. Sloughing of the under lip followed, accompanied by inflammation in-

volving the region of the lower jaw on both sides
of the face. Necrosis supervened, and considerable
portions of the jaw-bone were cast off. The condition of his face consequent thereupon was as follows:
The entire under lip was gone, and the skin below it.
On the right side of the chin the loss of skin extended to within one finger's breadth of the edge of
the jaw, while on the left side it did not extend so
low down. The upper lip was of ample dimensions,
and both angles of the mouth were entire and somewhat drawn down. The right half of the tongue
adhered by its under surface to the floor of the mouth,
as far forward as the alveolar margin of the maxilla
in front; its extremity, which was also bound down
by adhesions, was exposed to view from the absence
of the under lip. That portion of the body of the
lower jaw between the symphysis and last molar
tooth on the right side had previously come away
entire, and the alveolar sockets of all the teeth belonging to it could be identified in the bony specimen (*see* Fig. 21). The last molar tooth on the
right side alone remained *in situ*. A much smaller
portion had also come away from the left half of the
jaw. It consisted of the alveolar border supporting
the sockets of the two bicuspids and two adjacent
molar teeth (*see* Fig. 22). This extensive loss of bone
on the right side had been supplied by new bony
product along the entire inferior border of the maxil-

la to such a degree that the symmetrical form of the
face was maintained and a solid support afforded.
The canine tooth on the left side of the lower jaw
and the last molar on the right side were the only

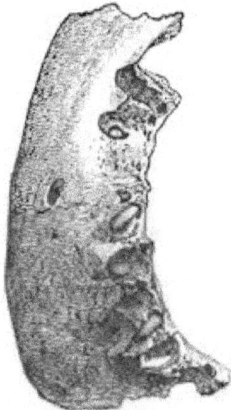 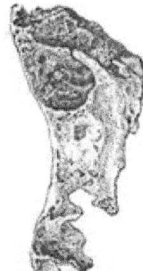

Fig. 21.   Fig. 22.

teeth remaining *in situ*.   Articulation was but little
affected.   The greatest discomfort experienced by the
patient was from the constant dribbling of saliva.
Owing to the loss of teeth his nourishment was
restricted to liquid and soft solid articles of food.
His general health was good.   (*See* Fig. 23, taken
from a plaster cast.)

First Operation. — Was performed October 5,
1869, at patient's residence, with the aid of Prof.
A. C. Post and Drs. C. M. Bell and J. N. Beekman.
It was executed as follows: The right cheek was
dissected up on the inside of the mouth, from the
jaw downward as far as the lower border of the

jawbone, and backward to a point a little beyond
the anterior edge of the masseter muscle. An incision was then carried from a point, *a*, below the middle of the zygoma downward upon the right cheek,
and forward in a curved line to a point, *b*, half an
inch below the right angle of the mouth. After dis-

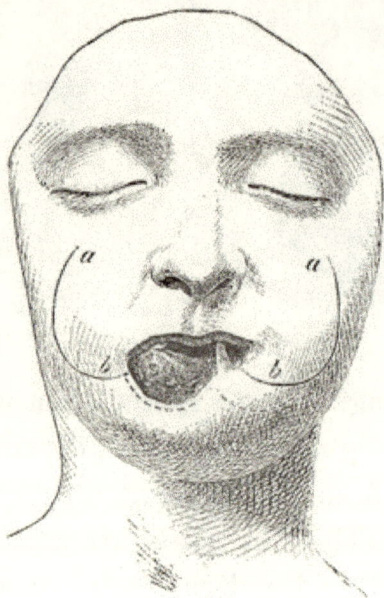

Fig. 23.

secting up the flap thus formed, and getting access
to the cavity of the mouth, the mucous membrane
alone was divided along the anterior margin of the
masseter muscle upward, and then forward as far as
the upper canine tooth. By this procedure the entire cheek flap was liberated, and could be advanced
edgewise forward till its free extremity, carrying

with it the angle of the mouth, reached the line of the symphysis menti. The same procedure was executed on the left cheek. The arteries were ligated as fast as they were encountered, thus sparing hæmorrhage. The skin covering the prominence of the chin was shaped symmetrically into an angular form, the angle pointing upward, and occupying the median line. The two cheek flaps were then glided edgewise forward, and made to meet by their anterior edges over the symphysis, where they were secured together by two pin sutures, inserted below the lip border, and three fine-thread sutures at the border. Below their line of junction over the symphysis the edges of the two cheekflaps diverged from each other, and stood astride of the angle of skin covering the chin, and were adjusted to it by sutures. The effect of this adjustment was to draw the angles of the mouth toward each other, and fold the upper lip upon itself, so that it stood forward in advance of and overhung the short, retracted under lip. The surfaces left bare on both cheeks by the transfer of the flaps, as just described, were covered by dissecting up from their underlying connections the edges of the wound from which the flaps had just been detached, and gliding them forward to meet the cheek flaps, and be again adjusted to them by sutures. By this means, at the completion of the operation, there remained no uncovered, bare surface. Notwithstand-

6

ing the extensive dissections required in this operation, the hæmorrhage was controlled within moderate limits, and was not followed by any marked depression of the pulse. A single attack of vomiting,
which emptied the stomach of the blood swallowed
during the operation, was the only disturbance produced by the protracted administration of ether.
The parts were covered with a layer of double thickness of sheet-lint to maintain their natural warmth.
The inflammatory tumefaction developed during the
three or four days succeeding the operation was moderate, as was also the febrile reaction. On the second
day I began removing the alternate thread sutures,
and changing the yarn on the pins. Special care was
required to maintain in close contact the edges of the
flaps that were united over the symphysis and on both
sides of the chin, so as to prevent, if possible, the
escape of saliva between the sutures. On the right
side of the chin, where the tongue came in contact
with the parts on the inside of the mouth, and exerted some pressure against them, saliva did escape,
and prevented adhesion from taking place. At all
other points, however, primary union was secured,
but not without renewing the sutures at different
points, whenever ulceration began to take place, before
the support of the original sutures could be dispensed
with. The administration of nourishment was man-
aged by patient's mother, whose long experience en-

abled her to do it very successfully. Union having failed to take place on the right side of the chin, there remained a narrow aperture of nearly two inches in length, with cicatrized, rounded edges, that required an operation for its closure.

Second Operation.—Performed October 30th, as follows: Both edges of the aperture were pared afresh. The lower edge was dissected up from the bone and everted; the upper edge was cut across at both ends, so as to permit it also to be everted. The two edges were then accurately confronted and secured together by six pin sutures and nine intermediate silver-wire sutures, the wound thereby being rendered impermeable to saliva.

November 4th.—The last suture was removed and union obtained at all points, except one, where an opening of the size of a goose-quill allowed the saliva to escape from the mouth. Under the repeated application of nitrate of silver and the support afforded by adhesive plaster, this opening was at length permanently closed on November 15th. The reconstructed mouth, as shown by Fig. 24 (taken from a

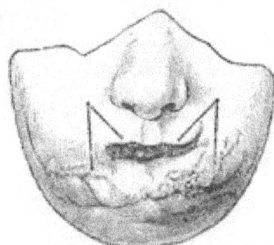

Fig. 24.

photograph of a plaster cast of the face), presents the upper lip as doubled up on itself, much increased in thickness, protruding, and consequently overhanging the short, tense, retracted under lip. To equalize the dimensions of the two lips, and restore the mouth to its natural angular shape, was the object of the next operation, which was performed on November 30th.

Third Operation.—The method employed was that described on page 22, and was applied first to the left half of the mouth only. A satisfactory result was obtained, and the natural angular shape of the mouth was restored on that side.

Fourth Operation.—On January 4, 1870, the same operation was applied to the right half of the mouth as had been applied to the left. It was feared that sloughing might take place on this side of the face from the presence of numerous cicatricial lines, resulting from the previous operations. None, however, occurred to mar the result of the operation. The apex of the triangular patch, after its transfer from the cheek to the side of the chin, did slough, but without any detriment to the result. By these two operations the mouth was restored to its natural shape and functions, except that there remained a notch on the right half of the under-lip border, near the angle of the mouth, where the saliva escaped uncontrolled. This defect in the result of the last operation, as shown in Fig. 25 (taken from a plaster cast), was owing partly to a greater deficiency of material on the right side of the chin originally, and partly to the right half of the under lip having united at the symphysis, below the level of the left half, after the first operation. The patient spent the following summer in the country, and returned to

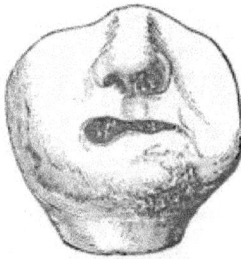

Fig. 25.

the city in October, much improved in health. The parts involved in the previous operations had also improved, having regained a good degree of softness and pliability. The discomfort caused by the escape of saliva demanded an operation for the closure of the notch upon the under lip. It was performed on October 5, 1870, as follows:

Fifth Operation.—The border of the notch was split lengthwise by an incision along its middle, and the edges on both sides of the split were pared off, beveling, so as to increase the thickness of the fresh-cut border. The upper lip was then transfixed a little above the vermilion border, at the middle of

its right half, and an incision carried toward and
around the right angle of the mouth in such a manner as to detach a strip of lip border, half an inch
wide, that would include the angle of the mouth.
This strip, still connected with the upper lip, was
brought round lengthwise, and secured to the fresh-
cut edge of the notch by fine-thread sutures, inserted close together. Though a material improvement was effected by this operation, the escape of
saliva was not entirely controlled, and the mouth
was somewhat shortened.

Sixth Operation.—Was performed on April 12,
1871, for the purpose of restoring the angular shape
of the mouth on the right side. It was executed
according to the method described on page 28. A
good result followed, and the natural angular shape
of the mouth was restored.

Again the patient spent the summer in the country, and on his return to the city, in September, a
further improvement in the condition of his face was
observable. The mouth, though restored to its natural angular shape on the right side, was still scanty
in length. A shallow notch on the right half of the
under-lip border also remained, and allowed some
escape of saliva, especially when the head was inclined forward, and his attention was not directed to
controlling it. After a little persuasion, seconded
by his mother's influence, the patient consented to a

final operation for remedying the defects still existing. It was performed on September 23, 1871, as follows:

Seventh Operation.—The first step of the operation was to lengthen the mouth on the right side by extending the angle further outward. This was accomplished by repeating the sixth operation, as above

FIG. 26.—Condition in March, 1872.

described. The next step was to obliterate the notch on the right half of the under-lip border. To accomplish this the following method was executed: Two incisions, commencing one on either side of the notch, a little below the lip border, and extending through the entire thickness of the lip, were carried downward in converging lines that met below the chin. The included triangular patch, having for its base the notched portion of the lip border, and retaining its connection for support on both sides of the notch, was pushed upward, and secured on a higher level by bringing together the edges of the wound below the patch, and securing them in contact by a pin suture, wound with cotton yarn. A second pin suture was inserted higher up, and made to traverse both edges of the wound, as well as the patch interposed between them; an additional thread suture below completed the adjustment. Primary union without any suppuration followed, and on the third day the last suture was removed.

As a final result of these several operations, the patient's mouth was restored to its normal shape and dimensions. The saliva no longer escaped uncontrolled. The lips performed their natural motions, and articulation was but little affected. Fig. 26, representing his condition in March, 1872, was engraved from a photograph. At the present time—March, 1873—still further improvement is noticeable. The numerous cicatricial lines, intersecting the surface on both sides of the chin and cheeks, have shrunk to the level of the adjacent surface, and they, as well as the skin itself, are perfectly soft and pliable.

Patient was again seen in January, 1876, and the good results of surgical treatment previously ascertained were found to continue unimpaired, as well in respect of the appearance of the parts themselves as of their functions.

CASE V. — *Reconstruction of the Mouth after Removal of the Under Lip for Disease.*

S. D., aged forty-nine, married, native of United States, a calker by occupation, was admitted into St. Luke's Hospital May 28, 1870, with epithelial cancer of the under lip. The narrative of the case will be restricted to the patient's condition at the time of entering the hospital, which was as follows: A salient, morbid growth involved the middle portions of the under-lip border to within half an inch of both angles of the mouth, and extended downward upon the chin to within a finger's breadth of its lower border. The left half of the tumor stood out prominently, and its everted margin overhung the adjacent sound surface. The upper lip was somewhat folded upon itself and drawn down at the angles of the mouth, in consequence of the removal of a V-shaped portion of the under lip by an operation performed in the month of January preceding, for the removal of a diseased growth involving the lip border, which had preceded the present growth. The wound had been treated after the operation in the usual manner, but before healing was complete the growth reappeared in the wound, and at length attained the dimensions above described. A glandular enlargement, about the size of a cranberry, existed under the jaw on the right side, at a point below the anterior edge of the masseter muscle. There was also a much smaller one on the left side, in a corresponding situation. (*See* Fig. 27, taken from a photograph of a plaster cast of the patient's face.)

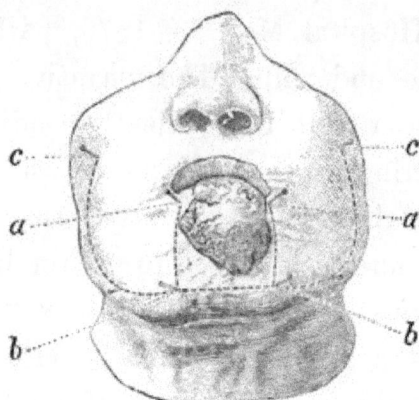

Fig. 27.

First Operation.—Was performed May 29th, as follows: An incision, commencing at a point within less than half an inch of each angle of the mouth, was made to divide the under-lip border, and extend vertically downward on either side of the morbid growth, till they both joined a transverse incision crossing the lower part of the chin, *a* to *b; a* to *b,* Fig. 27. The quadrilateral patch thus formed, in-

cluding the morbid growth, was dissected off from
the underlying periosteum. The transverse incision
was then continued outward on both cheeks to a
point within a finger's breadth of the angles of the
jaw, and thence extended upward a distance of two
inches in a line curving slightly forward (b to c;
b to c). The cheek flaps thus designed were dissected
up on both sides from their underlying connections,
and the mucous membrane, lining their inner surface,
was alone divided along the anterior margin of the
masseter muscle upward, and thence forward along
the line of its reflection, from the upper jaw to the
cheek, as far as the upper canine tooth. The two
cheek flaps, thus freely liberated, were glided forward
edgewise toward each other, and made to meet over
the symphysis menti, where they were secured in accurate coaptation by three pin sutures and intermediate fine-thread sutures. The spaces left bare on
the cheeks by the transfer just described were closed
by approximating their opposite edges, and securing
them together by sutures, which, owing to the laxity
of the skin, was effected without any dissection.
The facial artery was unavoidably divided on both
sides, and consequently had to be ligated. These,
with other ligatures that had to be applied to bleeding vessels, were gathered into a loose skein and
brought out at the nearest point of exit. The mouth,
as the result of this new adjustment, assumed a cir-

cular and pouting shape, which, however, still permitted liquid nourishment to be taken through a tube and by means of a spoon. In the subsequent progress of the case a moderate degree of inflammatory tumefaction supervened. Saliva escaped for a few days from between the edges of the wound on the right side of the chin. The edges of the cheek flaps, where they met in front over the symphysis menti, and were subjected to a great degree of tension, required the most careful management to secure final union. Fortunately union was secured by primary adhesion at the lip border, but it failed below to the extent of an inch, and here the first pins inserted had to be reënforced by a succession of fresh pins, and at last by resorting to a beaded-wire clamp suture, inserted in the manner described on page 17; and this first clamp was succeeded by a second one, which was finally removed and dispensed with on June 14th, when union was completed at all points. Fig. 28, which shows the result of this first operation, was taken from a photograph of a plaster cast.

Second Operation.—Was performed on the 23d of June, for the purpose of equalizing the length of both lips, and restoring the natural angular shape of the mouth. It was executed according to the method described on page 22, and was applied to both sides of the face at one operation. Fig. 28 shows with greater exactness the direction of the in-

cisions, as designated by pins inserted erect at selected
points and dotted lines drawn between them. For
the first few hours after the operation the surface of
the transferred patches had a pale, ashy look, which,
however, disappeared on the day following without
any subsequent sloughing. Primary union took

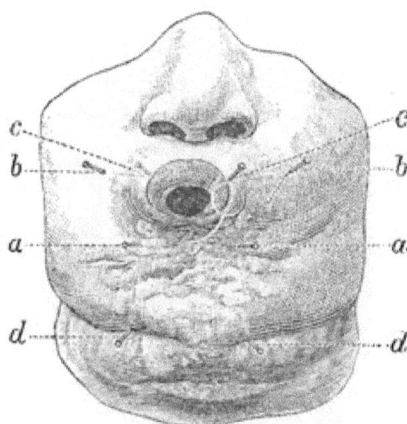

FIG. 23.

place at all essential points. The reconstructed
mouth presented a symmetrical and natural shape,
without any interruption of the continuity of the
lip border at its new angles. Patient was discharged
July 7th, at his own request, and readmitted October
14th, for the removal of a tumor situated in the submaxillary region of both sides; their existence had
been recognized before the first operation. They
were then of small size, and it was hoped they might
disappear after the removal of the diseased under lip.

They had, however, considerably increased in size,
and their true character was no longer doubtful.
Patient's general health had much improved. His
mouth retained its symmetrical and natural shape,
both angles being perfect in form. Though somewhat scanty in length, the mouth performed all its
functions so satisfactorily that patient was unwilling

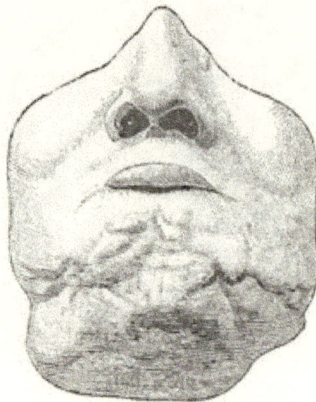

Fig. 29.

to accede to the proposal to increase its length by
another operation. The tumor on the right side adhered closely to the lower edge of the jaw, and after
removal was found of the size of a hickory-nut.
That on the left side was loose and movable, and of
the size of a cranberry. Both wounds having nearly
healed, patient left the hospital, and returned to his
home October 21st.

Fig. 29, also taken from a photograph of a plaster
cast, shows the final result. The author learned that

the disease reappeared several months subsequent
to his leaving the hospital, and progressed to a fatal
termination.

_____

CASE VI. — *Reconstruction of the Mouth after the
Loss of the Right Half of both Lips; also, Esmarch's Operation for Anchylosis of the Jaw.*

GEORGE K., aged six and a half years, of German
parentage, resident of Williamsburgh, Long Island,
N. Y., of fair complexion and light hair, was admitted into St. Luke's Hospital in May, 1866. The loss
of parts sustained in this case appeared, from the
father's statement, to have been caused by cancrum
oris, that occurred in the progress of typhoid fever;
whether it followed the administration of mercury
or not could not be satisfactorily determined. His
condition, when I first saw him, was as follows: One-
half of the upper, and two-fifths of the under lip,
with the right angle of the mouth and the adjacent
portion of the cheek, had been destroyed, leaving the
subjacent teeth and gum-surface exposed. The cicatrized margin of the cheek was retracted, depressed,
and closely adherent to the upper and lower jaws,
binding them together immovably in close contact.
The lining mucous membrane of the right cheek having been destroyed, the cavity of the cheek was oblit-

erated. What remained of the upper-lip border terminated below the septum nasi in a rounded, shrunken, and somewhat everted extremity. The remains of the under lip constituted about three-fifths of its original size, and terminated in a rounded, everted extremity at a point immediately below the right canine tooth of the lower jaw. The columna nasi had been destroyed, leaving the cicatrized inferior border of the septum exposed. In consequence of the closure of the jaw the introduction of food into the mouth could only take place along the cavity of the left cheek, and across the space between the last molar teeth and the ascending ramus of the jaw. Soon after his admission to the hospital a necrosed portion of the lower jaw on the right side was removed. It was found to consist of the entire breadth of the jaw vertically, and included three-quarters of an inch in length of its lower margin, while upon its upper margin there still remained the entire alveolar socket of the second bicuspid tooth, with one-half of the socket of the first bicuspid anterior to it, and one-half the socket of the first molar posterior to it. Notwithstanding the loss of so considerable a portion of the bone, the reproduction of new bone was so complete that no trace of deficiency could be detected by the finger passed over the inferior margin of the jaw. Patient's general health, though pretty good when he entered the hospital, steadily improved

afterward by the aid of generous diet and daily out-
of-door exposure.   (*See* Fig. 30.)

First Operation. — Was performed June 20th.
The right cheek was detached from the jaws above
and below, and the dissection continued in every

Fig. 30.

direction, till the jaws could be separated far enough
to admit the thumb edgewise between the front
teeth.   The thinned cicatricial edge of the right
cheek, bordering on the region of the angle of the
mouth, was pared afresh for adjustment to the new
lips. What remained of both lips was now detached from their connections, the upper lip by an

7

incision commencing at the inferior border of the ala
nasi, and dividing the entire thickness of the lip on
a line parallel with the lip border, and continued
onward to the middle of the left cheek. The lower
lip was detached by an incision across the middle of
the chin, continued onward, parallel with and as far
into the left cheek as the previous incision of the
upper lip. The bifurcated quadrilateral flap thus
obtained, which consisted of the remains of both lips,
and was lined with mucous membrane, was advanced
toward the right side of the face, and its two extremities adjusted, with the lip borders in contact
with each other, to the edge of the right cheek, already prepared. Pin sutures were employed to secure the ends of the flap in place, and thread sutures,
inserted close together, served to secure the upper
and lower edges of the flap in their new relations.
This adjustment was completed in such a manner as
not to have any strain upon the sutures. Warm-
water dressings were directed to be applied to the
parts, and liquid nourishment to be given through a
tube. Moderate inflammatory swelling followed the
operation, but it began to subside on the third day.
On the fifth day most of the sutures had been removed, primary union having taken place at nearly
all points. Patient's appetite was good and his general condition satisfactory.

July 5th.—Healing was complete at all points.

Means were employed after the operation to prevent the closure of the jaws, by keeping wedges of wood between the teeth during the progress of cicatrization. They could, however, be borne only a part of . the time, and ultimately proved of no avail. The newly-constructed mouth, as shown by Fig. 31, is

FIG. 31.

scanty in length, and situated mostly to the right of the median line, with its left angle on a line below the left orifice of the nostril. This defect, in the length and symmetry of the mouth, it was proposed to remedy by a second operation, which was performed on the following 23d September.

Second Operation.—Was executed according to the method described on page 28, and promised well. On the fourth day after the operation the last suture was removed. Though a considerable improvement in symmetry was effected by this operation, the mouth was still deficient in length, and scarcely exceeded one inch and a half. In the month of May, 1867, patient, after a long absence at home, was re-admitted into the hospital for a

Third Operation.—This was a repetition of the preceding one, and was performed for the purpose of extending the mouth still further at the left angle. The improvement in the symmetry of the mouth and the general expression of the face, resulting from this operation, was quite satisfactory. Patient still suffered, however, all the discomfort connected with the closed condition of his jaws, and could introduce food into his mouth only between the left cheek and teeth, as already described. The only expedient adapted to remedy this condition was Esmarch's operation for establishing an artificial articulation, at a point anterior to the cicatricial bands which held the jaws in contact. This was resorted to on July 1, 1867.

Fourth Operation.—It will be borne in mind that the cicatricial band holding the jaws in contact occupied the entire region of the right cheek, and terminated anteriorly in a callous edge, extending between the upper and lower bicuspid teeth. At a point anterior to this callous edge a portion of the lower jaw was to be excised, for the purpose of establishing an artificial articulation. All the muscles concerned in depressing the lower jaw, having their insertion on either side of the symphysis, would be undisturbed, and consequently would continue to perform their office in acting upon the jaw after the proposed operation. The procedure was as follows: An incision was made along the inferior edge of the jaw down to the bone, from a point near the angle to a point three-quarters of an inch distant from the symphysis. The outer and inner surfaces of the jaw were denuded to the same extent. The bone was then perforated by a drill of the size of a goose-quill on a line below the first bicuspid tooth, to facilitate the division afterward of the entire bone by the action of a strong, cutting bone-pliers. The same procedure was applied posteriorly on a line below the second molar tooth, and the included fragment of bone, measuring more than one inch in length, was removed. Special care was taken to avoid the facial artery by drawing it out of the way posteriorly. A mass of callous tissue, in which the upper teeth were imbedded, was pared away. The cut ends of the bone were gnawed smooth with Lüer's rongeur forceps. The anterior liberated portion of the lower jaw could now be separated from the upper jaw so as to admit a finger edgewise between the molar teeth on the left side. A tent of lint, of the size of the little finger, was inserted, with one end passing out at the right angle of the mouth, and the other through the wound below the jaw. The remainder

Fig. 32.

of the wound was closed by sutures. Water dressings were applied to the face and neck. The subsequent inflammatory action was moderate, and soon began to subside. The final result of this operation was as follows: The space left after the removal of the portion of jaw-bone diminished by the gradual approximation of the opposite ends of the bone.

The degree of mobility, however, of the anterior fragment of the jaw was only limited, but yet it sufficed to permit the introduction of food into the mouth directly between the front teeth. (*See* Fig. 32.)

The author regrets that he has not been able to discover the whereabouts of this patient, so as to ascertain his condition after the test of a long interval of time.

CASE VII.—*Mutilation and Distortion of the Mouth from a Shell-Wound.*

EGBERT H., aged twenty-two, native of Vermont, a private soldier in Company C, Sixth Vermont Regiment, Sixth Army Corps, received a wound on September 19, 1864, at Winchester, Va., from a fragment of a shell, that laid open the right cheek from the angle of the jaw to the mouth, lacerated the under lip at the right angle of the mouth, and carried away the upper and lower front teeth. The nose and upper lip were also split vertically, and the under jaw was fractured at the symphysis, with comminution of the bone. The parts having been healed for some time, the condition of the face, at the time of the operation, was as follows: The mouth was shrunk and contracted to half its natural size, and both lips shortened. The right half of the under-lip border overlapped and adhered to the alveolar border of the jaw, from which the teeth were absent. A deep notch in the lip, capable of lodging the little finger, had resulted from this adhesion, and permitted a constant escape of saliva from the mouth, to the great annoyance and discomfort of the patient. The surface of the skin bordering on the upper lip and right angle of the mouth was wrinkled and cicatricial. A conspicuous scar crossed the right cheek from the angle of the jaw to the mouth. The adhesion of the right half of the under lip to the jaw, though it involved the entire depth of the lip, did not spread out laterally. The loss of bone, consequent on the comminuted fracture of the lower jaw at its symphysis, had been followed by an approximation of the two halves of the jaw to each other, thereby narrowing the dental arch so that the upper and lower teeth of both sides could not be brought simultaneously into contact with each other. Owing to this deformity mastication and articulation were somewhat affected. Patient wore habitually a small tin gutter, to catch the constant flow of saliva. (*See* Fig. 33.)

First Operation.—Was performed October 28, 1864, at the Central Park Hospital, at the request of Surgeon Clements, U. S. A., in charge, and in the presence and with the aid of the medical officers of the hospital, as follows: Two incisions, dividing the under-lip border, were commenced, one at the right angle of the mouth, *d*, the other at the middle of the lip, *b*, and continued downward, in gradually-converging lines, to a point, *c*, under the chin. The V-shaped patch thus formed, including as it did the notch upon the lip border and the adherent portion

FIG. 33.

of the lip below it, was dissected off from the underlying periosteum. The remaining left half of the lip and the adjacent cheek were detached from the jaw as low down as its inferior border, and as far back as the last molar tooth, after the buccal mucous membrane had been first divided to the same extent

along the line of its reflection from the jaw to the
cheek. This procedure permitted the detached parts
to be glided toward the right side of the chin, and
thus to contribute to close the space left by the removal of the V-shaped patch of lip. In order to
obtain additional material for the same purpose on
the right side of the chin, a transverse incision, d to
a, was carried through the right angle of the mouth
across the cheek to within a finger's breadth of the
anterior border of the masseter muscle, and thence a
second incision, a to e, was extended downward and
a little forward to a point below the edge of the jaw,
on the same level with the apex of the V-shaped
bare space under the chin. The quadrilateral patch
thus formed, having its upper half lined with mucous
membrane, was dissected up from the jaw, but left
attached below it. It was then glided forward edgewise till it met the left half of the under lip, and
their confronted edges could be adjusted to each
other by pin sutures, and fine-thread sutures in the
intervals between the pin sutures. By this adjustment the deficiency of the under lip was supplied,
and the lining mucous membrane of the quadrilateral patch made to confront the surface of the jaw,
denuded of its mucous membrane by the previous
removal of the adherent portion of the lip. New
lip border was constructed on the upper edge of the
transferred cheek patch by excising a prism-shaped

strip of tissue from between the skin and mucous membrane, and, after lapping the latter over the former, securing them accurately together by fine sutures, inserted close to each other. As it was necessary to lengthen the mouth on the right side beyond its previous limits, an addition of three-fourths of an inch had to be made to the upper-lip border, which was done in the manner just described in its application to the under lip. A new angle was also constructed for the mouth by securing the opposite edges of the divided cheek together at a point where the newly-constructed upper and lower lip borders terminated. The reconstruction of the mouth being completed, there still remained a triangular-shaped space of raw surface left bare upon the right cheek by the transfer of the quadrilateral patch. To provide a covering for this space the transverse incision, which had divided the cheek to within a finger's breadth of the edge of the masseter muscle, was continued in the same line through the skin alone to the distance of one inch, and the skin bounding the raw surface posteriorly was then dissected up and stretched forward so as to meet the edge of the advanced cheek patch, and thus cover the bare surface. This adjustment was secured by the requisite number of thread sutures. The operation thus completed occupied three hours, including the interruptions necessary to maintain the anæsthesia. The inflammatory tumefaction succeeding the operation was considerable for the first three days, and then began to abate; the febrile reaction was moderate. Suppuration took place only at the lower angle of the wound under the chin, and was of short duration. The management of the sutures was the same as has already been sufficiently noticed. Toward the end of the second week patient left his bed and went about the ward. As a result of the operation patient's appearance was much improved; the saliva no longer escaped from the mouth uncontrolled, and both mastication and articulation were benefited. Certain defects, however, still remained: the mouth was one-sided, three-fifths of its length being situated to the right of the median line, and two-fifths to the left. The lips being without support, in consequence of the absence of front teeth, retracted toward the cavity of the mouth. Patient being anxious to have these defects remedied, if possible, engaged to return to the city for another operation, after a visit to his family. On his return to the city he was admitted into the New York Hospital, December 12, 1865.

Preliminary to a second operation, Mr. J. A. Bishop, a skillful dentist, of No. 34 East Twenty-first Street, kindly undertook to adapt to the patient's mouth a plate, made of vulcanite, with artificial teeth in front, to supply the deficient teeth and afford support to the lips. Mr. Bishop's success was complete, and the result satisfactory.

A Second Operation.—Was performed January 9, 1866, for the purpose of lengthening the mouth at the left angle, and thereby restoring its symmetrical shape. It was executed according to the method described on page 28. A good result followed, alike satisfactory to the surgeon and patient. Fig. 34,

Fig. 34.

which shows the final result, was engraved from a photograph taken before patient's discharge from the hospital, January 19, 1866.

CASE VIII.—*Loss of a Large Portion of the Lower Jaw-bone, with Extensive Mutilation of the Face and Distortion of the Mouth, produced by a Shell-Wound.*

WM. SIMMONS, aged twenty-one, a private soldier of Company 1, New York Heavy Artillery, was admitted into New York Hospital, October 26, 1865. On the 25th of March preceding, while serving with the army before Petersburg, Va., patient was struck on the right side of the face by a fragment of shell, that lacerated the cheek extensively, and carried away the body of the lower jaw. The condition of the face, after the injured parts had been healed for two months, was as follows: The body of the inferior maxilla having been carried away, the chin, for lack of support, had lost its prominence and was sunk. The entire right ramus of the jaw, including the angle and alveoli supporting the last two molar teeth, remained *in situ*, while on the left side the upper half of the ramus only was preserved. From the middle of the zygoma on the right side a linear cicatrice extended obliquely downward and forward upon the cheek to a deep depression at the angle of the mouth, where the extremity of the upper lip and neighboring parts adhered closely to the underlying alveolar border of the upper maxilla, from which the teeth had been carried away. The upper-lip border

retained its normal horizontal position, but its right
half was increased in length, in consequence of the adhesions at its extremity. The right half of the under
lip, having been completely detached from its connections by a lacerated wound crossing the middle of the
chin transversely, had, in contracting new adhesions,
settled below its proper level, leaving a space of one
finger's breadth between the right halves of the two
lips, from which saliva escaped constantly uncontrolled, to the great discomfort of the patient. The
surface of the left side of the chin was uneven and
intersected by cicatricial lines. On the inside of the
mouth, behind the under lip, a callous ridge stretched
obliquely across the floor of the mouth, and terminated on the left side at, and adhered to, the extremity of what remained of the ramus of the jaw.
This band, which formed a part of the cicatrice uniting the under lip to the chin, appeared to serve as
a substitute for the lost bone in front, and afforded
a firm support for the anterior attachments of the
tongue. The last upper molar tooth alone remained
in situ in the upper jaw on the right side, while the
canine and all its fellows posterior to it remained *in
situ* on the left side. By the aid of the finger, introduced into the mouth, the under surface of the right
half of the tongue was found adherent to the opposite surface of the floor of the mouth, and the pro-
trusion of the tongue was thereby rendered impossi-

ble. His food was restricted to soft solid and liquid
articles of nourishment. Deglutition was unimpaired,
but articulation was so defective that he was averse
to holding conversation, and preferred to communi-

FIG. 35.

cate with others by signs and the use of a pencil and
paper. His appearance was ruddy, and his general
health good. (Fig. 35.)

Operation.—Performed November 7th, as follows:
The under lip was detached by a transverse incision
carried across the chin above the cicatricial line connecting the lip with the chin, to a point one finger's
breadth below the left angle of the mouth. The lip

border, which before being detached had assumed a
fan-shape, could now be straightened out and applied
to the upper lip throughout its entire length. In
order to reconstruct the right angle of the mouth,
the upper lip was detached at its right extremity
from its adhesions to the upper jaw. A point was
then chosen on its border, at such a distance from
the median line as would make the two halves of the
lip equal in length, and at this point the border was
pared away obliquely. At a corresponding opposite
point upon the under lip the border was also prepared in the same manner, and the two lips were
then confronted and secured together at their fresh-
cut edges by a single pin suture and additional thread
sutures. The next step of the operation was to cover
up the depressed cicatricial line upon the cheek which
was so conspicuous. It was accomplished by carrying two parallel incisions, one on either side of the
cicatricial line, downward and forward to the angle
of the mouth. The edges of these two incisions were
then dissected up, and brought together and secured
by sutures, so as to cover up the cicatrix, after its
surface had first been pared and made raw. This
adjustment was extended to the newly-reconstructed
right angle of the mouth. The same procedure was
then applied to the chin. An incision was carried
across the chin below and parallel with the cicatricial line already mentioned, to a point below the

8

left angle of the mouth. The lower edge of this transverse incision was next to be adjusted to the inferior cut edge of the detached under lip, so as to cover up the cicatrix intervening between them. In order to effect this adjustment the lower edge of the transverse incision required to be everted, which could only be effected after carrying a vertical incision from the terminus of the transverse incision below the left angle of the mouth, a distance of one inch and a half upon the neck. In the course of this vertical incision a flattened cyst, of the size of a silver half-dollar piece, filled with brownish viscid fluid, was encountered under the skin, and removed entire. The angle of skin included between the last two incisions was dissected up, and its transverse edge made to overlap and cover the cicatricial line, and meet the inferior border of the detached under lip; the two were adjusted to each other by sutures. Pin sutures were employed in this adjustment at selected points where the greatest support was required, and between them thread sutures were added. The reconstructed mouth had now assumed its normal shape and dimensions with both lips in contact. No adhesive plaster was applied. Warm-water dressings were directed to be kept upon the part, and liquid nourishment allowed to be given through a tube.

8th. — Progress favorable. Inflammatory tumefaction not excessive; febrile reaction moderate.

Changed the yarn on the pins, and removed some of the alternate thread sutures.

9th.—Doing well. Changed yarn; removed additional sutures, where they could be dispensed with.

10th.—Primary union secured at nearly all points. Removed all the pins and most of the remaining

Fig. 36.

thread sutures. A free suppuration escaped from the lower angle of the wound, below the chin. A superficial slough formed over the right zygoma, but it in no way marred the result of the operation. Strips of adhesive plaster were now applied at different points to support the newly-healed parts. The suppuration gradually diminished, and at length ceased, and healing was finally completed at all points. The lips, restored to their natural relations to each other, performed their functions, and controlled the escape of saliva. Some improvement in articulation and a marked improvement of the general expression of his countenance were observable. On the 12th of December following patient returned to his home in the country. In May, 1866, he revisited the hospital. A further improvement had taken place in his general appearance, and especially in his articulation, which had now become so much more distinct that he no longer shrunk from engaging in conversation, having discarded entirely the use of pencil and paper in communicating with others. The result is shown by Fig. 36.

CASE IX.—*Closure of an Opening into the Superior Meatus of the Right Nasal Fossa.*

MARGARET K., aged twenty-one, native of Ireland, unmarried, was admitted March 3, 1869, into the New York Hospital, and gave the following statement of her case: After passing through an attack of scarlet fever when nine years old, a black spot appeared on the right side of the upper part of her nose, and was the occasion of the opening which still exists. Its situation and shape are accurately represented in the accompanying Fig. 37, taken from a photograph. It is of an ovoid form, more than one inch in its vertical diameter, and three-quarters of an inch in its greatest transverse diameter. It communicates directly with the right nasal fossa, and allows a free passage to the air. Its outer edge approaches close to the inner canthus of the right eye.

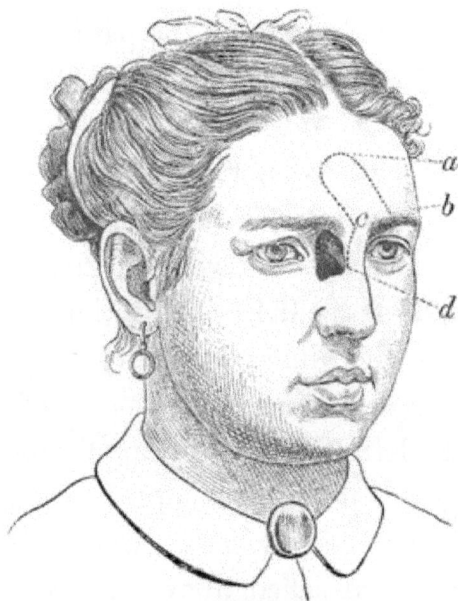

FIG. 37.

The skin surrounding it is thin, pliable, and somewhat overlaps its margin; the inner surface of the cavity, as far as it is visible, appears healthy. Her voice is unaffected. A patch of adhesive plaster is habitually kept applied to the opening to conceal it. Patient being naturally anxious to be rid of so conspicuous a disfigurement, consented willingly to an operation for its removal.

Operation.— Performed March 5th, as follows: The skin at the margin of the opening was dissected up and everted, a procedure requiring great care, owing to the thinness of the skin, especially where it bordered on the inner canthus of the eye. A pattern of the size of the opening, cut from oiled-silk, served as a guide in shaping the patch of skin which was to be raised from the forehead. To prevent this patch from becoming too much twisted, when transferred to its new locality, its position on the forehead was so chosen that its long axis inclined obliquely at an angle of forty-five degrees toward the opening, and its pedicle of attachment lay above and close to the inner extremity of the left eyebrow. After dissecting up the patch of skin designated by the letters b, a, c, a strip of skin, e, d, intervening between it and the opening, was also dissected up, to make room for the transfer of the patch, but left attached above the right eyebrow, and reserved for subsequent use. After adjusting the patch to the opening by sutures, the reserved strip of skin was utilized to cover the surface left bare upon the forehead, which it did so completely that no bare spot was left uncovered after the completion of the operation. Warm-water dressings were directed to be applied to the parts.

The subsequent progress of the case was favorable, and primary union followed at all points, except at that portion of the circumference of the patch which bordered on the inner canthus of the eye; here union failed to take place, in consequence of the opposite edges becoming inverted. To remedy this defect a second operation was performed April 6th.

Second Operation. — The ununited edges were pared afresh, and two pin sutures inserted, special care being taken to evert the edges and maintain

FIG. 38.

their cut surfaces in contact. Additional thread sutures were also necessary to complete the adjustment. Perfect union followed this operation. The transplanted patch, from some excess in size, formed in its new locality a rather conspicuous bulging ridge, particularly between the eyebrows, where its pedicle had undergone a twist. To remove this disfigurement a third operation was performed April 17th.

Third Operation.—A prism-shaped strip of skin was excised from the ridge along its entire length, and the opposite edges of the wound brought together into exact coaptation, and secured with numerous fine-thread sutures. Primary union followed, and a level surface remained. The final result was a decided improvement in the appearance of the face. (*See* Fig. 38.)

CASE X.—*A Rhinoplastic Operation for the Restoration of the Apex Nasi after it had been bitten off.*

W. W. G., aged thirty-five, a resident of the city, was brutally assaulted in the evening of May 12, 1872, and during the affray had the apex of his nose bitten off by his assailant. The condition of the parts, twelve days after the occurrence, was as follows: All inflammatory swelling of the nose had subsided. A healthy suppurating surface indicated the extent of the lost parts, which included the skin covering the apex, and adjacent ridge of the nose as high up as its middle, and also both sides of the nose to a point within half an inch of the junction of the nose with the cheeks. The entire denuded surface was equivalent to about one-third of the superficies of the organ. The alæ nasi were disconnected anteriorly, and both had sustained about an equal amount

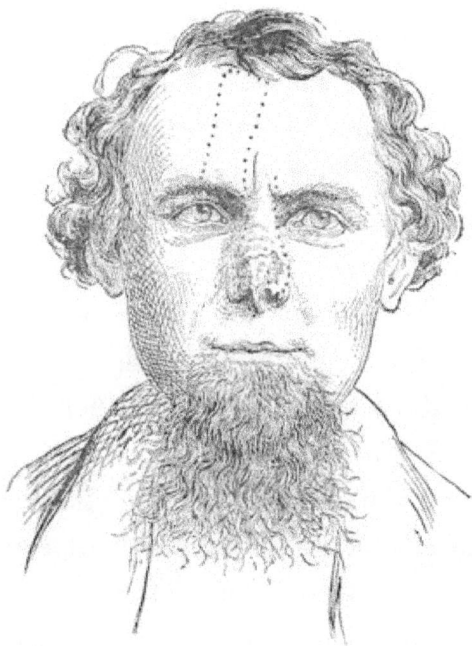

Fig. 39.

of loss of substance. The columna remained entire. The ridge of the cartilaginous septum was denuded, but had sustained no loss of substance. (*See* Fig. 39.)

Operation.—On Saturday, May 25th, an operation was performed at patient's residence, with the aid of Prof. A. C. Post, M. D., and Drs. T. E. Satterthwaite, J. N. Beekman, and N. S. Westcott, as fol-

lows: The anterior edges of what remained of both alæ
were pared and made straight, and an incision was
then carried upward on both sides of the nose, on a
line continuous with these edges, as high as the inner
extremities of both eyebrows. The patch of skin
included between these two incisions was dissected
up from the dorsum nasi and between the eyebrows,
and left connected above. A pattern, of the shape
of the entire denuded surface, cut from oiled-silk,
was laid upon the forehead in an inverted vertical position, immediately above the inner half of the right
eyebrow, with one edge bordering on the incision
that terminated at the inner end of the eyebrow.
The patch of skin underlying the pattern having
been outlined by an incision, was then dissected up
from the pericranium, but left connected below at
the margin of the orbit. Suitable allowance had to
be made for shrinkage of the patch, especially in the
direction of its length, after its transfer.

The skin, previously displaced from the nose and
from between the eyebrows, was now dissected up
still further toward the left side of the forehead, in
order to afford a sufficient extent of raw surface, with
which the under surface of the forehead patch might
come into immediate contact after being transferred
to its new locality. The patch of skin from the forehead was now brought around edgewise from right
to left, and from above downward, till it reached its

destination, and was spread out upon the nose, and made to extend downward beyond the apex and overhang it. Special care was taken that there should be no strain at the pedicle of the patch, where it was doubled upon itself, and where any obstruction of the circulation would endanger its vitality. In order to confront more exactly the edges of the patch with the opposite edges of the space it was to occupy, the latter were dissected up and everted sufficiently for the purpose. Pin sutures were inserted at select points, and fine-thread sutures in the spaces between them, to secure the adjustment. The patch of skin displaced from the nose and between the eyebrows was now carried upward, and employed to cover the surface on the forehead left bare by the transfer of the forehead patch. It proved sufficient to fill up the lower half of the space, and was secured there by sutures. The remaining upper half of the bare surface, which encroached upon the hairy scalp, was covered with a collodion crust (*see* page 13). Both patches of skin, in being transferred, were necessarily doubled upon themselves, and formed two prominent folds of a flattened conical shape, standing out on the surface, one being situated above each eyebrow. (Fig. 40, *a* to *b*.) But few ligatures were required to secure bleeding vessels, and these were brought out at the nearest point of exit. Wet applications were avoided, and the parts covered with a double thickness of sheet-lint, spread on one side with cerate to prevent its adhering. The operation occupied two hours, and was well borne by the patient. His subsequent progress was favorable and requires no special notice. On the fourth day the last sutures were removed. On the sixth day the

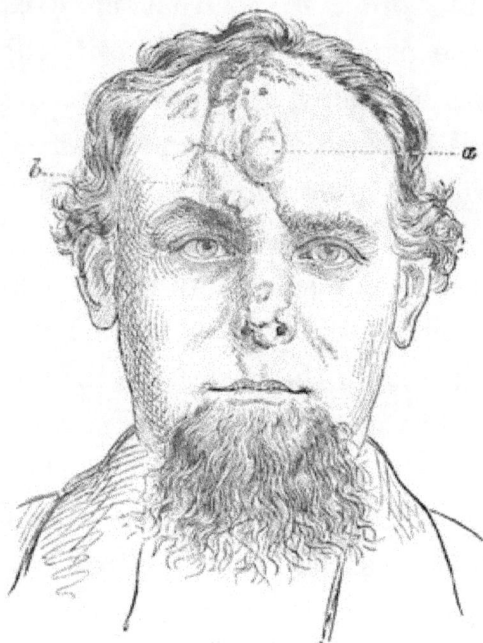

FIG. 40.

collodion crust became detached from the forehead and came off, and the underlying healthy, granulating surface was thereafter dressed in the ordinary way, till it finally cicatrized. The result of the operation is shown by Fig. 40.

Second Operation.—Was performed June 28th, for the purpose of leveling the two prominences re-

maining upon the forehead, one being situated over
each eyebrow. The procedure was as follows: A
curved incision was carried half around the base of
each prominence on its broadest side, and the prominence itself raised up from its underlying connections; and split across its middle. After being un-
folded and spread out, the redundant skin was pared

Fig. 41.

away, so as to be level with the surrounding surface.
after the edges had been adjusted to each other and
secured by sutures. This operation did well, and
was followed by the desired result. The end of the
patch of skin covering the nose, which overhung the

apex, was trimmed and shaped after having shrunk as much as it would. Fig. 41, showing the final result, is from a photograph, taken on the succeeding 12th of October, when the following particulars were noticed: Sensation, which for several weeks following the first operation, continued to be referred to the forehead when the surface of the patch on the nose was irritated, had regained its normal condition, and was referred to the actual seat of irritation. The principal cicatrice upon the forehead, where the patch of skin had been removed, had shrunk to a small size, and was concealed by the hair. The other cicatrices were only linear and not conspicuous. The lower third of the dorsal surface of the nose having been derived from the scalp, continued to yield its hairy growth, and required to be shaved at short intervals. Patient was seen as late as October 14, 1875, when all the parts involved in the operation above reported continued in good condition.

SECOND CLASS.

By far the most numerous of this class are cases of harelip. Their successful treatment depends so much on the proper execution of the several details which make up the operation, such as the management of the patient, the preparation of the parts for readjustment, the choice of sutures, the manner of inserting them and their subsequent management till healing is completed, that, in order to avoid repetition, some general remarks on these several topics will first be presented.

1. *The Management of the Patient.*—In the case of a child the arms should be brought down to the sides of the body, and secured by a folded napkin passed around them and fastened with pins. The head should be held steadily in position between the hands of an assistant. In the administration of an anæsthetic, after its full influence has been produced, its continued action during the progress of the operation may be most conveniently maintained by the

use of a small sponge, held in a dressing forceps, and applied to the nostrils of the patient.

2. *Preparation of the Parts for Readjustment.*— In order to hold the parts on the stretch, and thereby enable the operator to make the requisite incisions with the utmost precision, the following expedient will be found very satisfactory: A needle, armed with a coarse thread, is to be passed through each half of the lip at the angle bounding the cleft on either side, a point being chosen for the passage of the needle that will not interfere with the subsequent incisions. The ends of each of the threads when tied together will form loops with which to hold the parts on the stretch when required. Each half of the lip is then in turn to be held on the stretch and drawn away from the jaw, while the lip and cheek are detached from the upper jaw by an incision of the buccal mucous membrane, carried along the line of its reflection from the jaw to the lip and cheek as far back as the molar teeth, if necessary. The dissection is also to be extended upward toward the orbit on the level of the periosteum. This procedure, after being applied on both sides, will allow the two halves of the lip to be approximated, and the confronted edges of the cleft to be secured in contact, without there being any strain upon the sutures that are to hold them together. The opposite edges of the cleft are next to be pre-

pared as follows: Each half of the lip, being again held on the stretch, is to be transfixed near its angle by a Beer's cornea knife, and an incision carried upward, along the border of the cleft, as high as, and if necessary into, the nostril. The strips thus detached from both borders of the cleft, but left attached at the angles below, are to be brought down with their fresh-cut surfaces facing each other, and both transfixed with a threaded needle. The ends of the thread tied together afford a loop with which the strips are held evenly on the stretch, and the fresh-pared edges of the cleft at the same time made to confront each other, while one pin suture is inserted close to the vermilion border below, and another close to the columna nasi above. Two or three thread sutures are to be inserted between the pin sutures. The two detached strips of cleft border, still connected at the angles, are now to be severed by dividing them, not at right angles across their length, but obliquely, so that, after their cut surfaces are confronted and united by three fine-thread sutures, they will form a projection standing out beyond the line of the lip border. By this latter expedient (which is credited to Malgaigne) the formation of a notch at the point on the lip border where the two halves unite is best prevented. The advantage of employing Beer's cornea knife, as recommended above, is, that with it the lip can be more

9

readily transfixed and the sections completed with greater precision. When harelip is complicated with

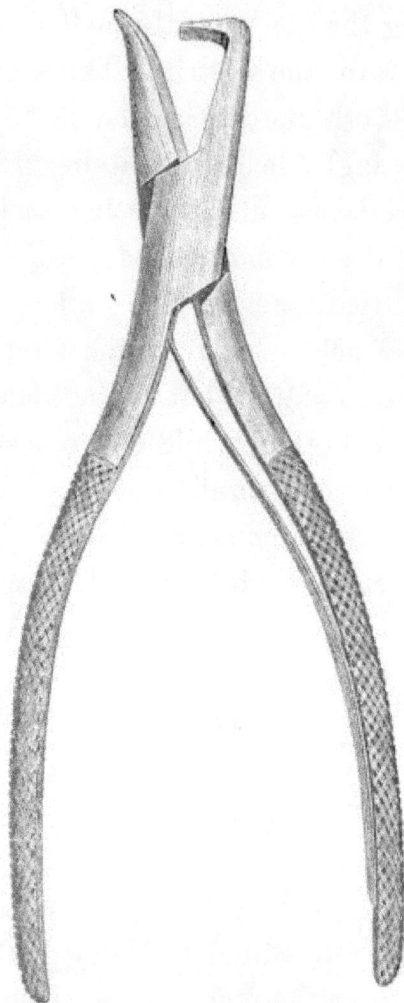

Fig. 42.

a cleft of the dental arch, the cleft divides the arch into two segments of unequal length, and the anterior extremity of the longer segment projects in advance of the natural curve of the arch, forming a prominence which interferes with the approximation of the two halves of the lip. This prominence requires to be broken down and reduced into position, an operation best effected by means of Butcher's (of Dublin) bone-pliers, an instrument devised for the purpose (*see* Fig. 42). In employing it the bony prominence is to be seized somewhat crosswise between the two blades, with the bent blade applied upon the anterior surface, high up toward the nostril. On pressing the blades together, the end of the bent blade sinks into the bone, and, while the instrument is acted upon by means of the handles from before backward, a fracture is produced and the prominence reduced into its place, where it fills up the notch, and may be made to adhere permanently, if the opposite edges have been previously pared and made raw for the purpose. Consolidation follows without exfoliation of bone or suppuration.

3. *Sutures and their Management.*—This subject has already been fully treated in an introductory chapter, to which the reader is referred.

HARELIP.

The following classification embraces all the varieties of this deformity, of which examples are to be found in the subsequent pages.

First Class.—Single cleft of the upper lip alone, involving most frequently its left half, immediately below the nasal orifice.

Second Class.—Double cleft of the lip alone, involving sometimes only its border; more frequently, however, its entire vertical dimensions.

Third Class.—Single cleft of the lip, complicated sometimes with cleft of the dental arch alone; more frequently with cleft of the bony and soft palates, in addition to the dental arch.

Fourth Class.—Double cleft of the lip, complicated with cleft of the bony and soft palates, and the coexistence of an intermaxillary bone.

To avoid repetition, it may be here stated that all the cases of harelip about to be narrated were operated on according to the method described above. A rare instance of the hereditary development of harelip will be first reported.

A Harelip Family.—Mrs. Molinieri and her three children, all girls, natives of Genoa, Italy, were admitted into St. Luke's Hospital, January 10, 1871. Mrs. M. herself bore the marks of a successful opera-

tion for harelip performed in childhood. She had a
brother and sister with harelip; and, besides her
three living children, she had had four others, who
had all died in early infancy. Three of them had
harelip, and the fourth one only was a perfect child.
In other words, there were nine instances of the deformity in two generations of a single family.

Case XI.—*Single Harelip.*

Marie Anne, the second child of this family,
aged four, had a single cleft, involving the left half
of the lip. The left border of the cleft was vertical

Fig. 43.

in its direction; the right border slanting, and diverging from the left (*see* Fig. 43). Operated on
January 21st. Primary union followed, and the last

suture was removed on the fourth day after the
operation. Strips of adhesive plaster were continued a few days longer, to support the new ad-

FIG. 44.

hesions and relieve them of all strain. Fig. 44
shows the result, without any notch remaining at
the lip border.

CASE XII.—*Double Harelip.*

ROSE, the infant, twelve months old, with double
cleft of the lip, separated by a central tongue-shaped
portion. The clefts involved only about three-fifths
of the depth of the lip, and did not quite extend
to the orifices of the nostrils above (*see* Fig. 45).

Operated on January 21st. The borders of the central and lateral portions of the lip, after having been pared, were adjusted together and secured by a pin, traversing all the three portions at their upper part, and by fine-thread sutures in addition. Some difficulty was encountered in reducing the central piece

Fig. 45.

to the same level with the lateral pieces, and was not entirely overcome. Primary union failed to take place, except at the vermilion border. By the employment of adhesive plaster, renewed daily, with great care, union by granulation was at length obtained at all points, and completed on the twelfth day. The elevation of the central piece above the level of the lateral pieces persisted to a slight degree, but might be expected with time gradually to

FIG. 46.

diminish. Fig. 46 shows the result, with only a slight notch remaining at the lip border.

CASE XIII.—*Single Harelip, with Cleft of the Dental Arch and Bony and Soft Palates.*

JACINTA, the eldest child, aged seven. In this case the cleft of the lip involved its left half; the right and largest segment of the alveolar arch terminated anteriorly, at the margin of the cleft, in a rounded prominence, which stood forward in advance of the curve of the arch, and, being uncovered by the

lip, formed a conspicuous feature in the disfigurement
of the face. The left ala nasi, being unsupported,
was retracted toward the cheek, so as to flatten the
nose on its left side (*see* Fig. 47). Operated on Jan-

FIG. 47.

uary 31st. The bony prominence formed by the anterior extremity of the right segment of the alveolar
arch was first broken down and reduced into position by the application of Butcher's bone-pliers, as
described on page 129. The prominence, after it was
reduced, bridged over and filled up the cleft in the
alveolar arch. By previously paring the confronting
edges bony consolidation was secured. The removal
of this prominence also facilitated the approximation
of the two halves of the lip and their adjustment to
each other. Primary union was secured only at the

lip border. It therefore became necessary to hold
the two halves of the lip in contact while union by
granulation was taking place. This was effected by
the employment of a beaded silver-wire suture, which
was inserted high up toward the nose, and at a distance of nearly one inch on either side of the confronted edges of the cleft. The clamp suture was left
in situ five days, and healing was complete at all

Fig. 48.

points at the end of three weeks. A tooth, growing
out of the bony projection that had been reduced,
became loose and was extracted. The fractured
parts became consolidated without exfoliation or
suppuration. A plug of sponge was worn in the left
nostril, to keep it distended and in good shape. No
notch remained at the lip border. Fig. 48 shows the
result.

Other examples of single harelip uncomplicated:

Case XIV.—*Single Harelip on the Right Side.*

John Black, aged twelve, from Piermont, on the Hudson, entered St. Luke's Hospital, January 18, 1871. The cleft divided the right half of the lip, and extended upward, as a shallow furrow, along the floor of the right nostril. The two halves of the

Fig. 49.

lip were but little separated from each other. The right middle incisor tooth occupied the cleft, and stood out conspicuously in advance of its fellows, owing probably to the lack of support from the lip (Fig. 49). Operation performed January 24th, and followed by primary union. The last sutures were removed on the fourth day. Adhesive straps were renewed daily for a week longer, when healing was

Fig. 50.

complete. A very slight notch remained at the lip border. Fig. 50 shows the result.

Case XV.—*Single Harelip.*

A female infant, five months old, from Hudson, Columbia County, N.Y. The cleft, as usual, involved the left half of the lip (*see* Fig. 51). The operation was performed October 12, 1870, and followed by primary union, without any remaining notch at the lip border. She returned to her home October 30th. Fig. 52 shows the result.

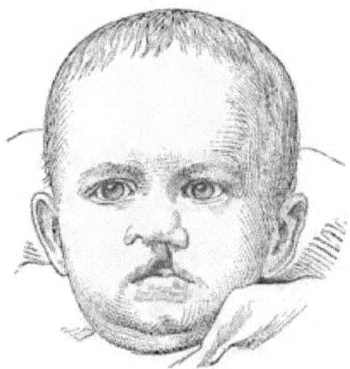

Fig. 51.　　　　　　　　　　Fig. 52.

Case XVI.—*Single Harelip.*

A MALE infant, three months old, from Hudson City, N. J., with single cleft of left half of the lip (*see* Fig. 53). Operated on June 11, 1873. Primary

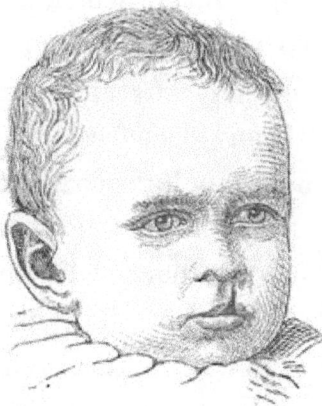

Fig. 53.　　　　　　　　　　Fig. 54.

union followed. Resumed nursing on the seventh day; returned home on the tenth. No notch remaining at the lip border. Fig. 54 is from a photograph taken in August following.

CASE XVII.—*Double Harelip, complicated with Cleft of the Bony and Soft Palates, together with the Presence of an Intermaxillary Bone.*

HARRIET Q., aged eight, from Winsted, Conn., entered New York Hospital, March 16, 1865, with the above deformity of congenital origin. A wide cleft divided, in an antero-posterior direction, the dental arch and bony and soft palates into equal halves. The bony septum narium traversed the entire length of the cleft in the median line. Its inferior border was thin and rounded posteriorly, but expanded out anteriorly into an irregular-shaped mass, which constituted the middle portion of the alveolar arch, and supported the two middle and left lateral incisor teeth, standing out irregularly from its surface. This projecting bony mass, known as an intermaxillary bone, was somewhat deflected from the median plane to the right, so that its right margin touched the anterior extremity of the right segment of the dental arch, thereby obstructing the view into the cavity of the nostril of that side, while it exposed more fully to view the left nasal cavity. The upper lip was cleft below both nasal orifices, and the two halves of the lip on either side, with their attached alæ nasi, yawned wide apart, thereby increasing the breadth of the nose, and giving it a very flattened appearance. A central tongue-shaped portion of the upper

FIG. 55.

lip was interposed between the two clefts, and rested upon the surface of the projecting intermaxillary bone. It was continuous with, and, as it were, suspended from the columna, which was scarcely half an inch in length. The characteristic defect in articulation which usually attends this condition existed in a marked degree. (*See* Fig. 55.)

Operation.—Performed March 18th, as follows: The projecting intermaxillary bone was excised with bone-forceps, on a line with the inferior border of the septum, after first dissecting up from its upper surface the tongue-shaped portion of lip, as high as its junction with the columna, to which it was left attached. This latter was then brought down into contact with the fresh-cut edge of the septum, and adapted to it by squaring its edges and securing them by sutures to the mucous membrane on both sides of the septum. By this means the scanty columna was lengthened out to its full dimensions. The two halves of the lip were then detached extensively upward and outward, on both sides, from the anterior surface of the upper maxilla, to facilitate their approximation to each other in the median line. In order still further to facilitate their approximation, and also to narrow the cleft in the dental arch, and at the same time restore its natural curve, the projecting ends of its two lateral segments, bordering the cleft on either side, were broken down and reduced into position by the application of Butcher's bone-pliers, as described on page 129. The opposite edges of the cleft in the lip were then prepared and adjusted to each other by sutures, in the manner described on page 126. After the completion of the operation, warm-water dressings were directed to be applied to the parts, and appropriate liquid nourishment, with stimulants in moderate quantity, to be given. The subsequent progress and management require no detailed description. On March 5th the last sutures were removed, and adhesive-plaster applied to support the newly-united parts.

April 29th.—Patient returned to her home in the country. The only remaining incisor tooth in the

FIG. 56.

upper jaw had to be extracted, owing to ulceration of the opposite surface of the upper lip from pressure against it. As a final result of the operation, the dental arch in front was restored to its natural curve, and bony consolidation took place without exfoliation or suppuration. The cleft involving the alveolar arch anteriorly had become narrower. A

10

slight notch remained at the lip border, where the
two halves had united. Fig. 56 is from a photograph taken more than one year after the operation.

––––––––

THE following are examples of cases of harelip,
operated on a second time for the purpose of remedying the imperfect results of previous operations
performed in infancy.

CASE XVIII.—A daughter of Mrs. S., aged six,
from Cold Water, Mich., had been operated on, in

FIG. 57.

infancy, for single cleft of the right half of the upper

lip. As the result was very imperfect, the parents were anxious to have the defect remedied, if possible, by a second operation. (*See* Fig. 57.)

Operation.—Performed October 12, 1871. A good result followed after primary union, and without any remaining notch at the lip border. Healing was complete on the sixth day. The result was especial-

Fig. 58.

ly gratifying, from the fact that other surgeons, whom the parents had consulted, had counseled against an operation. One year later the mother wrote: "Hardly any trace now remains of the scar, and we have no doubt she will entirely outgrow it." Fig. 58 shows the result.

CASE XIX.—A physician's daughter, aged sixteen, from Savannah, Mo., had been operated on, when five months old, for single harelip, involving the right half of the lip, with a very unsatisfactory result, as shown by Fig. 59. In order to remedy the

FIG. 59.

defect, a second operation was performed May 27, 1873, and followed by primary union and complete healing on the sixth day. On the tenth day she left for home with her father.

Under date of August 23d following, her father, in a letter inclosing her photograph, from which Fig. 60 was taken, writes, "The condition of her mouth is still improving."

Fig. 60.

CASE XX.—Miss C., of New York, aged thirty, was operated on, when seven months old, for single harelip, complicated with cleft of the bony and soft palates, involving the left nasal cavity, but without extending through the dental arch anteriorly. The restoration of the upper lip alone was attempted, but the result of the operation was unsatisfactory. A conspicuous notch remained at the border of the lip, and the union between the two halves of the lip was incomplete at the upper part, so that the outer orifice of the left nostril extended considerably below the level of that of the right. The dental arch, not being involved in the cleft, retained its natural curve. (*See* Fig. 61.)

Fig. 61.

Operation, *May* 14, 1874.—After separating the two halves of the lip with scissors, applied exactly on the line of the cicatrice uniting them, the fresh-cut edges were prepared for readjustment to each other by the same method as would be employed in a first operation for harelip (*see* page 126). Special care was requisite to secure adhesion of the opposite edges of the cleft at its highest point, where it involved the orifice and floor of the nostril, and where union had failed to take place after the first operation. After the opposite edges of the cleft within the orifice of the nostril had been pared and made raw, a beaded silver-wire suture was employed for the purpose of maintaining them in close contact after they were confronted. The wire was made to traverse the confronted raw edges, and was armed at one end with a bead, resting against the septum nasi, low down in the right nostril, and at the other end with another bead, resting against the left cheek, where it joins the ala nasi. The adjustment of the two halves of the lip below and at its border was completed in the manner already described (*see* page 127). The details of the subsequent progress and treatment it is unnecessary to relate, except that the beaded suture was left *in situ* till the eighth day, and performed a useful service in securing adhesion at the point where it was most difficult to effect it, and where we most feared that it would fail. Prior to the operation just described there existed a want of symmetry in the mouth. The portion situated on the right side of the median plane exceeded in length that on the left side. The difference was somewhat increased, and became more conspicuous, after the second operation. In opening her mouth patient herself experienced resistance to the separation of the lips on the left side. To remedy this defect she consented to another operation, the design of which was to lengthen the mouth on the left side by extending the angle toward the cheek. It was performed June 19th, according to the method described on page 28. Although an abscess formed in the left cheek, adjoining

FIG. 62.

the angle of the mouth, and a small slough was cast off, the final result was satisfactory, and the symmetry of the mouth was restored. In the month of October following, Mr. J. A. Bishop, a skillful dentist of this city, undertook to adapt a plate of vulcanite to the roof of the mouth, that would close the cleft in the bony palate, and also supply four artificial

teeth, that were to take the place of the four upper
front incisors, which it was thought best to extract,
on account of the irregular position they occupied.
Mr. B.'s success was complete. The plate, when finally adapted, was worn with much comfort, and the
characteristic defect of articulation, which had existed in a marked degree, was almost entirely cor-
rected. Fig. 62, from a photograph taken in April,
1875, shows the final result.

CASE XXI.—*Congenital Hypertrophy of the Tongue.*

MARY JANE C., aged nine, from Birmingham, Conn.,
of a healthy family and good constitution, entered
St. Luke's Hospital, May 21, 1866, with hypertrophy
of the tongue, which, her mother stated, had existed
from birth, and had progressively increased with her
general growth. The condition of the tongue at the
time of her admission was as follows: It protruded
habitually from the mouth about two inches beyond
the lips, and distended the angles of the mouth, and
was overlapped by them. It measured one inch in
thickness, two inches and a half in breadth, and four
inches and five-eighths in circumference. Its upper
surface was coated with a yellowish-brown crust,
which on drying became detached and fell off in
scales. The protruded portion was of firm consist-

ence and painless. The portion within the line of the
teeth, though of full size, was in other respects normal. The under lip hung over upon the chin from
the pressure of the tongue. The tongue could be
protruded about four inches beyond the front teeth.
The lower front teeth were depressed nearly to an
horizontal position, and thickly incrusted with tartar,

Fig. 63.

so as to be more than double their normal size. This
was true of all the lower incisors and both canine
teeth. It was remarkable how little her articulation
was affected. She attended school, and was accustomed to recite and sing with her fellow-pupils. The
sublingual and submaxillary glands were not enlarged. Her general health and aspect were good.
(See Fig. 63.)

First Operation. — Performed May 26th. The head of the patient was supported against the breast of an assistant, who retracted the angles of the mouth with his forefingers. The tongue was then seized with a volsella, and drawn forward out of the mouth. A strong ligature was then passed from below upward, through both edges of the tongue, as far back as possible. The ends of each ligature being tied together, formed loops with which to control the tongue and draw it forward. While held forward and spread out laterally by means of the loops, the tongue was transfixed with a straight, sharp-pointed knife through its middle, from below upward, as far back as the teeth would permit, and a flap formed on the left side by cutting forward and outward. The right flap was formed by applying the knife to the opposite edge of the tongue, and cutting in a reversed direction inward and backward. The arteries spurted briskly, but were readily secured in the usual manner and ligated. The flaps were then brought into contact and secured by sutures. It was now for the first time noticed that the lower jawbone itself had undergone a change of form in front, in consequence of the constant pressure of the tongue. The teeth of the upper and lower jaws came in contact only as far forward as the first molars on both sides, while the front teeth were more than one inch apart at the symphysis.

May 27*th.*—Patient obtained some sleep; pulse 126; febrile reaction considerable. Ordered tinct. aconit. rad. gutt. j; spirit. Mindereri, ℈ij, q. 4. hor.

28*th.*—Slept well; pulse 120; swelling of tongue considerable; its extremity has a livid, ashy aspect; breath offensive; right side of the face swollen; deglutition somewhat difficult. Ordered wine, and a mouth-wash of dilute permanganate of potash.

29*th.*—The lateral flaps have separated, and their confronting surfaces are sloughy.

31*st.*—Swelling abating; sloughs separating; deglutition improving; general condition satisfactory. Subsequently no considerable loss of substance resulted from the sloughing. Healthy action was gradually reëstablished, and on June 18th cicatrization was complete. Although the excess in the breadth of the tongue had been reduced by the operation, there still remained an excess in its thickness at the end of the stump, which occupied the space between the front teeth, and could not be retracted. A second operation was therefore resorted to, to remedy this defect, and was performed on July 1st following.

Second Operation.—A strong ligature was passed transversely through the back part of the tongue by means of a long darning-needle, and the two ends made use of to hold the tongue forward. The tongue itself was seized with a volsella, applied at opposite points to its edges, near the extremity, and compressed it laterally, so as to increase its thickness. While thus held, a wedge-shaped portion was removed by transfixing the tongue laterally, at a point far back and equidistant from its upper and lower surfaces. The under flap was first formed by cutting forward and downward through its under surface, while the upper flap was formed by applying the edge of the knife upon the upper surface, at a point opposite to the extremity of the newly-formed under flap, and cutting in a reversed direction backward and downward to the point where the first section had been commenced. After ligating two or three arteries, the two flaps were brought into exact contact, and secured by interrupted thread sutures at the margin. By this second operation the tongue was reduced in thickness and length, so that it could now easily be retained within the limits of the dental arch, and its extremity had assumed a rounded, flattened shape. On the second day after the operation the tongue had become much swollen, and considerable febrile reaction had taken place; pulse 124; the surface of the body hot and dry. On the seventh day all the sutures had been removed in succession. There had been no sloughing, nor any separation of the flaps, as after the first operation. The swelling having mostly subsided, deglutition had become easy, and patient was able to be up most of the day. Convalescence progressed favorably till about the 10th of July, when she began to complain of sore throat, which was accompanied by fever, restlessness, loss of appetite, etc.

July 12*th.*—The submaxillary glands were hard and tender, and the whole submaxillary region much swollen externally.

16*th.*—The tongue itself was also much swollen; articulation had become very indistinct and deglutition very difficult; a copious secretion of saliva flowed constantly from the mouth; pulse 144. Ordered poultices of flaxseed-meal to the throat and a Dover's powder at bedtime.

19*th.*—The swelling has softened and fluctuation is perceptible. Subsequently a spontaneous opening formed under the tongue, and discharged pus abundantly into the mouth, affording great relief, and followed by an abatement of the swelling and a decided amelioration of all the symptoms. Recovery thenceforward progressed rapidly, and without further drawbacks. Before leaving the hospital, Mr. J. A. Bishop removed the tartar incrustation from the teeth, and adapted a fixture with elastic bands, with which to exert upward pressure under the chin, for the purpose of restoring the under jaw to its normal relations to the upper. The fixture was to be worn as much of the time as possible. She returned to her home, in Connecticut, on July 27th. In July of 1869 the author saw the patient, and on examination found her condition as follows: The tongue retained its normal dimensions, and occupied the floor of the mouth, its upper surface not rising above the level of the lower teeth. When the act of protruding the tongue was attempted, it bowed itself upward and

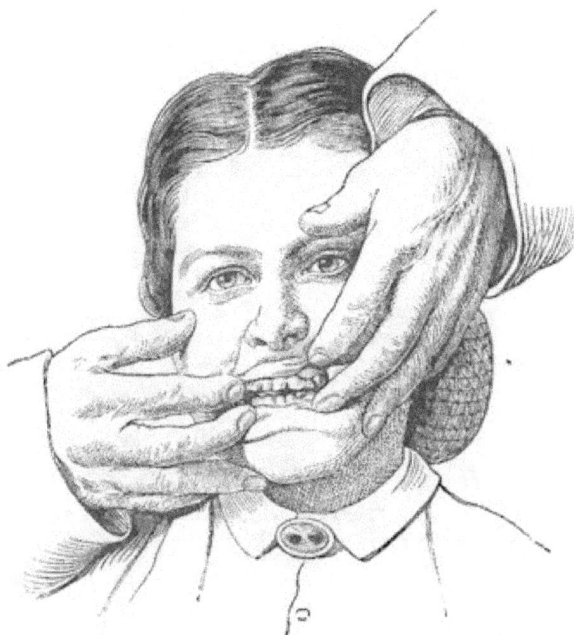

Fig. 64.

somewhat forward, being sustained by the connections of its under surface. The lower front teeth had regained their erect position, and the separation between the upper and lower teeth in front was scarcely half an inch. Articulation was without any defect. (See Figs. 64, 65, from photographs taken at that time.)

Remarks.—The method employed in the second operation that was performed in the above case is obviously the one to be preferred in similar cases of

Fig. 65.

hypertrophy of the tongue. It has the advantage of reducing more effectually the excess of thickness in the organ than the method employed in the first operation, and leaves the tongue of a more natural shape after the operation. It also avoids severing the connections of the muscles inserted into the under surface of the tongue, and, as the incisions do not involve the floor of the mouth, there is no communication opened between the wound and the intermuscular spaces in the submaxillary region, and consequently there is less liability to burrowing suppuration.

_____

CASE XXII.—*Congenital Hypertrophy of the Under Lip.*

THOMAS P. A., aged twenty-five, a native of England, a printer by trade, resident in the United States since three years old, was admitted into St. Luke's Hospital, February 5, 1867. From birth his lower lip had been abnormally large, and with it there co-existed a raspberry-colored stain of the surface of the chin and cheeks wherever they were covered by the beard. During youth and early manhood the parts underwent no special change other than an apparent increase of the swelling of the lip and a brighter redness of the neighboring surfaces in hot weather. While exposed to the hardships of military service in the far West, during the late war, a fresh impetus was given to the growth of the lip, and it increased to double its previous size. In August, 1866, successive operations, consisting of the introduction of red-hot needles, were performed by the late Dr. Charles A. Pope, of St. Louis. Sixty insertions in all were made at various intervals within a period of six weeks. Inflammation and slight suppuration followed; the parts became more dis-

11

tended, and the permanent result, as patient believes, was rather detrimental, the lip remaining larger than before, and retaining a sensation of numbness. He had also less control over its motions than before the operation. Nothing similar to his condition is known to have existed among his relatives. His condition at the time of admission into the hospital was as follows:

The raspberry discoloration occupied the surface on both sides of the face and chin, where the beard grows. The lower lip was more than double the thickness of the upper lip, and proportionately increased in all its other dimensions. It was pendulous, of a soft, flabby consistence, and free from pulsation. In its substance small, hard knots were felt, which had existed only since the insertion of the hot needles. Simultaneous compression of both common carotids had no perceptible effect on the volume of the lip. (*See* Fig. 66.)

Anticipating copious hæmorrhage in the performance of an operation upon the lip, clamps were devised for making compression at both angles of the mouth. They consisted of two flat steel blades, half an inch wide, bent flatwise at right angles, and made to slide lengthwise upon each other on their flattened surfaces by means of a screw. The vertical portion of the blades beyond the angle was two inches in length, and adapted to compress not only the lip proper, but also the skin below the lip as far down as the edge of the jaw. Before applying the clamp, an incision was made through the mucous membrane on the inside of the lip where it joins the bone, at a point below the angle of the mouth, to allow the end of the distal blade to be forced down in contact

Fig. 66.

with the periosteum as far as the edge of the jaw. The proximal blade was then screwed up, and the included lip and skin below it were tightly compressed between the two blades. Both clamps having been applied in the manner described, an operation was performed on February 27, 1867, as follows:

A letter-V-shaped patch, including about three-fifths of the lip border, equidistant on either side from the angles of the mouth, and having its apex low down in the median line under the chin, was removed. As one of the clamps became accidentally loosened without causing any considerable hæmorrhage, it was evident that the clamps might have been entirely dispensed with. A single artery only, and that at the lower part of the wound, required a ligature. After dividing the mucous membrane along the line of its reflection from the jaw on either side of the wound, the opposite edges of the wound were brought together, and secured in exact coaptation by five pin sutures, inserted at equal distances from each other, below the lip border. Between every two pin sutures a silver-wire suture was added. Three fine-thread sutures were inserted at the vermilion border of the lip, one of them on its buccal surface. No adhesive-plaster was used. In order to destroy the raspberry discoloration of the face, which caused such a conspicuous disfigurement, a disk-shaped cautery-iron, with a smooth, flattened face, heated to redness, was applied to that portion of the surface situated above the angle of the jaw on both sides, and the application was immediately followed by compresses wet in ice-water, and frequently renewed.

The wound did well. On the third day all the

sutures had been removed, and primary union had taken place at all points. Superficial eschars separated from the burnt surfaces, leaving a healthy granulating sore without any remaining raspberry discoloration. After an absence of three weeks from the hospital, patient was readmitted April 2d, to undergo a second operation. The lip, though much improved, still retained its original excess of thickness. The granulation growth upon the cheeks had become too exuberant, and required a single application of solid caustic potassa, after which the nitrate of silver sufficed to regulate it.

A Second Operation was performed April 10th, to reduce the excess in thickness of the under lip, as follows: A prism-shaped strip was excised from the entire length of the lip-border by two parallel incisions, including between them one-third of the thickness of the lip, and penetrating deep into its substance. The edges of the wound were then brought together, and secured by fine-thread sutures, inserted close together. This operation did well, and produced a still further improvement in the lip. Anxious to return home, patient was discharged April 30th. The experimental treatment of the discoloration of the face by the actual cautery had the desired effect of destroying the raspberry tint, but, in consequence of patient's absence during the healing process, the cicatricial surface, resulting from the burn, was not so smooth and even as it might have been under proper management.

Fig. 67, which shows the patient's present condition, is from a photograph taken in January, 1872, five years after the operations. Dr. I. D. Beebe, of

Fig. 67.

Hamilton, Madison County, N. Y., through whose kind services the photograph was obtained, in a letter to the author, says that, "were it not for the remaining discoloration of the skin, which is considerable, no one would imagine he had had an operation on the lower lip, as the scar shows but very little."

CASE XXIII.—*Abnormal Growth of Hair upon the Forehead.*

F. J., aged thirteen, a native and resident of the city, with fair complexion, light hair, and of a good constitution, has from birth had an abnormal growth of hair upon the left half of the forehead, covering the entire surface between the left eyebrow and the hair of the scalp, above as well as toward the left temple. The hair of the left eyebrow is much coarser than that of the right, while the abnormal growth of hair is short, soft, and fine, like that of a mouse-skin; both are darker colored than the hair of the scalp. The surface of the skin covered by this growth of hair, after being shaved clean, was found to be of a tawny hue, like that of the scrotum, and slightly elevated. (*See* Fig. 68.) Both the patient and his parents being anxious to have this conspicuous disfigurement remedied, its removal was undertaken and successfully accomplished by destroying the abnormal growth, partly by the actual cautery, but chiefly by the application of solid caustic potassa. The first application was made November 15, 1867. A disk-shaped cautery-iron, with smooth, flattened face, was heated to redness and applied over one-fourth of the shaved surface, and held in contact with it until a dark-brownish eschar was produced. Immediately thereafter compresses wet in ice-water were applied to the part and frequently renewed. By this means patient almost entirely escaped suffering on coming out of the anæsthesia. After the separation of the eschar the sore surface was dressed with simple cerate, and the granulation-growth kept level with the neighboring surface, during the subsequent cicatrization,

Fig. 68.

by the application of solid nitrate of silver, renewed every second or third day. On November 27th a first application of solid caustic potassa was made to another fourth of the surface, by rubbing the end of the stick of caustic into the surface till the skin became disorganized into a soft, pulpy paste. The action of the caustic was then neutralized with diluted vinegar. Two subsequent similar applications were required to complete the cauterization of the hairy surface. Nitrate of silver was also used to regulate the granulation-growth till cicatrization was completed. By the end of the following Decem-

FIG. 69.

ber the entire surface had healed. For more than two years afterward the patient remained under the author's observation. The cicatricial surface upon the forehead continued smooth and pliable, and could be moved freely over the underlying parts. It was a little paler than the adjacent skin. No contraction affecting the upper eyelid followed. The improvement resulting from the treatment was every way satisfactory. (*See* Fig. 69.)

Case XXIV.—*Erectile Tumor of Large Size.*

An infant daughter of Madame de F., six months old, and of good physical development, came under surgical treatment in January, 1871, for an erectile tumor of congenital origin, seated upon the right shoulder, and covering the upper half of the deltoid region and the adjacent acromion process above. Anteriorly it encroached somewhat upon the pectoral region. Its longest diameter was two inches and three-quarters, and its shortest two and three-eighths. The outer half of the tumor was the thickest, and its margin, which was elevated more than one inch above the neighboring skin, was somewhat shelving; the inner half grew thinner and sloping toward its margin, which was but little higher than the adjacent surface. The surface of the tumor was unbroken, and of a bright raspberry-color; no pulsation was perceptible on it, nor were there any enlarged blood-vessels visible on the surrounding surface (Fig. 70). The treatment consisted in the employment of the actual cautery and the solid stick of caustic potassa. A bullet-shaped cautery-iron,

heated to redness, was the instrument used. Patient
was anæsthetized with sulphuric ether on each occa-

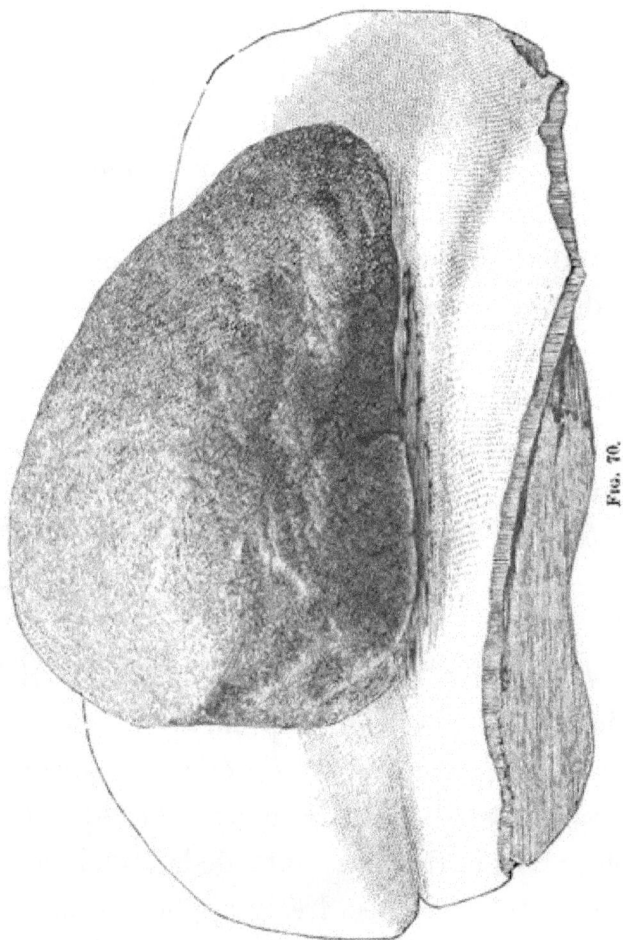

Fig. 70.

sion of the application of the actual cautery and
caustic.

January 14*th*.—A first application was made.
The iron, heated to redness, was sunk into the sub-

stance of the tumor, at one point in the centre and
at five or six points at its circumference, and made
to char the tissues with which it came into contact.
Immediately after the application, compresses wet in
ice-water were laid upon the part, and frequently
renewed for several hours afterward. The hæmorrhage was slight, and not sufficient to require attention. On emerging from the anæsthesia, patient did
not manifest any signs of severe suffering. A moderate degree of febrile reaction supervened on the
following day, but soon subsided thereafter; the local
inflammation was also moderate. After the separation of the eschars a dressing of simple cerate was
employed.

February 11*th.* — A second application of the
actual cautery was made in the same manner as the
first.

22*d.* — A first application of the solid stick of
caustic potassa was made by plunging it into the substance of the tumor, and then promptly absorbing the
fluids saturated with the caustic, in order to restrict
its action within the limits intended; and, in addition
to this, vinegar was also freely applied, to more completely neutralize the action of the caustic.

March 8*th.* — A second application of caustic was
made, and on March 22d a third and final application. The suppuration produced by these applications was at no time excessive, and the patient's

general condition was scarcely disturbed during the progress of the treatment. On the 12th of April following, when seen for the last time (in 1871),

Fig. 71.

there were no remains of the erectile tissue, the sore presented a healthy aspect, and was nearly healed.

On December 22, 1875, nearly five years after the treatment, the seat of the tumor was examined, and presented the following appearance: There was

no remaining discoloration, but a pale, smooth cicatricial surface in the place of the tumor. The skin
also was pliable, and moved freely on the underlying
parts.

Fig. 71, from a photograph taken at this time,
shows the condition of the shoulder.

CASE XXV. — *A Pendulous Tumor* (*Molluscum
Fibrosum*) *arising from the Right Half of the
Forehead and Temple.*

MISS R. R., aged thirty-five, a resident of the city,
came under surgical treatment on account of a pendulous tumor arising from the right half of the forehead and temple, and involving to some extent the
hairy scalp above. The tumor was elevated more
than an inch above the level of the surrounding surface, and in the progress of its gradual development
had gravitated toward the cheek and become invested
with the right eyelids, which were spread out upon
its surface. It had also dragged the eyeball itself
partly out of the orbit. The tarsal margin of the
upper lid, which overlapped and concealed the under
lid, formed the lower limit of the tumor, and descended upon the cheek to a point on a level with
the middle of the nose. The right eyebrow was
spread out upon the surface of the tumor one inch

below the level of the left. The skin covering the
tumor retained its natural color, was thinned, and
could be gathered into folds between the thumb
and fingers. Its substance was of moderately firm
consistence, but of loose texture, and the tumor itself glided freely upon the surface underlying it.

FIG. 72.

When grasped in the hand it had almost the pliability of the scrotum. On separating the lids, the eye
was found to be natural in appearance, and to possess good vision. (*See* Fig. 72.) When an infant a
few months old, her mother first noticed a somewhat
elevated ridge across the right half of the forehead,
which increased very gradually till it reached its
present dimensions. Its growth became accelerated

after her recovery from a dangerous illness in childhood, for the treatment of which she was profusely
salivated. When eleven years old, an operation was
performed, and a portion of skin of the size of a silver
quarter of a dollar excised, but without any benefit.
Other growths, varying in size from a small pea to a
cranberry, are scattered over the body, upon the limbs
and trunk. The patient allowed one of these growths,
of a pendulous form, near the fold of the right elbow,
to be removed for microscopic examination prior to
the performance of an operation on the face. Unfortunately, a memorandum of the result of the examination was lost. It was, however, ascertained that
there were no cancerous elements in the structure of
the tumor. Patient's general health being favorable,
the first operation was performed January 13th.

First Operation. — A transverse incision, commencing above the left eyebrow, was carried across
the tumor along the middle of the forehead to the
right temple; a second incision, commencing above
the right temple, was carried vertically downward
to the upper part of the cheek, joining the first incision at its terminus on the temple, the two incisions
forming a letter T laid upon its side. The angular
flaps of skin thus formed were dissected up from the
surface of the tumor to its margin above and below.
The tumor itself was then raised at its circumference
from the underlying surface, and a prolongation of

it, extending into the orbit under its roof, was dissected out. No large vessels were encountered, nor was the hæmorrhage from small vessels considerable. Ligatures were applied to all vessels requiring it. After the removal of the tumor the redundant skin was pared away, so as to permit a good adjustment of the flaps to each other. Numerous thread sutures were then inserted to secure the adjustment. A good result followed, with primary union at almost all points.

Second Operation.—April 8th. A flattened portion of the tumor, situated high up on the right temple, not having been included in the previous operation, a second operation, of only limited extent, was performed for its removal, and was followed by prompt healing of the wound.

Third Operation.—Performed April 28th. The right eyeball having been attacked with acute inflammation, which proceeded rapidly to disorganization of its internal structures, it was extirpated, and at the same time some remaining portions of the morbid growth were removed. A good result followed the operation, but the remaining conjunctival cavity unfortunately was not sufficiently capacious to permit the insertion of an artificial eye.

Patient was seen as late as September 11, 1873, and her condition ascertained to be as follows: She enjoyed ordinary good health. There had been no

return of the morbid growth; the cicatricial lines
upon the forehead and face had become quite indistinct; the right eyelids were habitually closed and
somewhat sunken. Fig. 73, from a photograph taken

FIG. 73.

in October, 1873, accurately represents the final result of the three operations.

Remarks.— In vol. xxxvii. of the "Medico-Chirurgical Transactions of 1854," the late Dr. Valen-
tine Mott reported five cases of what he termed "a
peculiar form of tumor of the skin," and denominated
"pachydermatocele." Two of the five patients were
boys, aged respectively twelve and fourteen years;
three were females, aged twenty-four, forty, and forty-five years.   In all the five cases the growths were

of congenital origin, or dated from very early childhood; they were all of a pendulous form and lax
texture. In the boys the tumors occupied the side
of the head and face. In two of the females the side
of the chest below the mamma was the seat of the
growth, and in the third the tumors occupied the
entire left side of the neck, between the larynx and
ligamentum nuchæ, and extended upward as high as
the ear, and downward upon the sternum, clavicle,
shoulder, and scapula; they hung pendulous as low
down as the umbilicus. All the patients were operated on for the removal of the tumors, and in all the
subsequent healing process advanced kindly. One
of the female patients, forty years old, died in a low
typhoid state within a few days after the operation,
her condition being induced by self-imposed priva-
tions of all sorts, from a fancied conception of extreme poverty. Three of the patients, one boy and
two females, recovered without any return of the
growth; one of these three, a female, forty-five years
old, whose case was the most formidable from the
number and large size of the tumors, withstood, in
the progress of her recovery after an operation, two
formidable attacks of erysipelas that endangered her
life. One of the four cases of recovery was a boy,
fourteen years old, on whom a second operation was
performed on account of the return of the growth.

Though both operations did well, the growth reappeared again after the second operation. In regard to the nature of one of the tumors examined, Dr. Mott states, " My colleague, Prof. Lovett, has kindly furnished me with the following notes of the microscopical appearance of one of the tumors: ' The specimen appears to me to consist of an hypertrophy of the skin and the subcutaneous cellular tissue. Under the microscope I find nothing but the exaggeration of the natural tissues; there are no evidences of a malignant formation.' " In vol. lvi. of the " Medico-Chirurgical Transactions," Mr. Pollock, of St. George's Hospital, London, has reported a case of what he terms "molluscum fibrosum, or fibroma." The patient was a widow, thirty-three years old, in whom the first appearance of the growth dated back to her earliest recollection. Besides numerous growths varying in size from a split pea to a small walnut, scattered over different parts of the body, a group of very large-sized tumors arose from the right side of the neck, between the hairy scalp and shoulder behind, and in front between the ear above and a line across the chest four inches below the top of the sternum. They hung in festoons over the right mamma, and reached below the umbilicus. An operation for their removal was followed by a favorable result. Four months after the operation there had been no return of the growths. Of the structure of these growths Dr. Whipham, after a microscopical examination, reported to Mr. Pollock: "The growth is due partly to an excessive hypertrophy of the connective tissue of the true skin, and partly to an abundant cell-growth occupying interspaces between the bands of fibrous and elastic tissue, which, as has been shown, comprise the chief part of the growth. Neither the large pendulous, nor the smaller sessile tumors, depend upon any alteration of the epidermis, rete-mucosum, glands, or hair-bulbs, as far as can be made out." In vol. xvi. of the "Transactions of the "Pathological Society," London, Dr. H. G. Wight reported the case of an unmarried woman, thirty-four years old, with an enormous group of pendulous tumors arising from the right half of the neck, between the symphysis menti and the spinous processes of the cervical vertebræ. They originated as low down as the fourth dorsal vertebra posteriorly, and anteriorly as low as a line drawn between the two nipples. Below this line they reached to the umbilicus. "The skin," he says, "resembles in some measure that of the scrotum. Patient first noticed the growth when about fourteen years old. She was found dead from suffocation, which, as was supposed, had occurred during an attack of epilepsy, to which she was subject." The structure of the tumors in all the above cases, including the author's, was strikingly the same, and consisted essentially of an overgrowth of the constituents of the subcutaneous connective tissue.

See also Virchow, "Die krankhaften Geschwülste," vol. i., pp. 350-353.

THIRD CLASS.

CICATRICIAL CONTRACTIONS FOLLOWING BURNS.

The cases belonging to this class, which have come under the author's own observation, will be arranged in two groups.

First Group. — Comprises lesions involving the face, and producing distortion of the eyelids, nose, and lips. Their treatment calls into requisition the resources of plastic surgery, and consists essentially in first liberating the distorted parts by detaching them from their underlying connections sufficiently to permit them to be restored to their normal relations, and then transplanting sound skin from the nearest available locality, with which to fill up the space made bare by the restoration.

Second Group.—Comprises lesions involving the neck and upper limbs, and producing contractions at their articulations, chiefly in the direction of flexion. In the cases belonging to this group, which are hereinafter narrated, the cicatricial bands producing the contractions consisted almost exclusively of integument, the underlying aponeurosis being involved only to a slight degree, while the muscles and tendons were not at all concerned. Their treatment consisted—

1. In excising the corded folds of cicatricial integument that maintained the contraction, and then dividing the edges of the resulting wound at every point where any resistance still remained, which prevented complete extension of the part. At some points it was also necessary to dissect up the edges from their underlying connections. In a word, the liberation of the parts required to be so thoroughly performed that they would have the utmost freedom of motion.

2. The liberation of the parts having been accomplished in the manner above described, the next step in the treatment was to adapt a suitable mechanical appliance, that would maintain the parts in their restored position during the process of cicatrization, and for a long time afterward. If the front of the neck should be the part involved, as in Case XXVIII., a brace would be required to keep the chin constantly elevated and the neck upon the stretch. The ingenuity of the surgeon must supply such modifications of the mechanical support as the exigencies of each case may require. It being necessary to wear the support constantly and for a long time, the comfort of the patient should never be lost sight of.

3. A third and very important step of the treatment is, to regulate the process of cicatrization during

its progress and until its final completion, in order that the healing surface may be kept smooth and even. To do this, two expedients are indispensable, and both of them require to be employed with unwearying perseverance. The constant tendency of the granulation-growth to become exuberant makes it necessary to repress it by the frequent application of caustics, of which solid nitrate of silver and caustic potassa have been preferred by the author. In applying the nitrate of silver, it is not sufficient simply to pass the stick over and in contact with the surface of the granulations, as is commonly done. It requires to be buried deeply into their substance at numerous points in close proximity to each other; and often the application has to be repeated at each daily dressing. Even this energetic use of the nitrate of silver will not always be sufficient to control the exuberant growth. It then becomes necessary, with the aid of anæsthesia, to have recourse to the solid caustic potassa, special precautions being taken in using it to restrict its action within the limits intended. This may be done by promptly absorbing with lint the fluids that so readily dissolve the caustic, and become saturated with it; and also by the application of common vinegar, which neutralizes its action. The potassa being much more energetic in its action than the nitrate, is only required at long intervals. In conjunction with the use of caustic, compression is an efficient and indispensable auxiliary, and is best employed by means of adhesive plaster, cut into strips, and applied in immediate contact with the entire granulating surface. The strips should overlap each other, like shingles on a roof, and they require to be renewed at each daily dressing. If in the progress of cicatrization there should be a tendency to the reproduction of contracting bands or folds, they should be freely divided with the knife at two or more points across the direction of their length. The formation of new skin-growth may further be promoted by leaving small patches or islets of the cicatricial skin at different points upon the raw surface, instead of removing it entirely. These islets will subsequently send forth from their edges in every direction new growth, and thus accelerate the healing process. It should be borne in mind that, for a long time after healing is completed, measures must be perseveringly employed to oppose the tendency to recontraction, which is very persistent; such measures, for instance, as wearing, during the night and a part of the day, the mechanical support that had been used in the earlier period of the treatment; also, the frequent stretching of the parts by manual force, together with the practice of appropriate gymnastic exercises. The practical application of these several methods of treatment will be best understood from the description of their use in the subsequent cases. These views of the treatment of cicatricial contractions agree in all important particulars with those taught by Dupuytren in his "Leçons Orales," vol. ii., Art. 1, p. 66, *et seq.*, Paris, 1832. The late Mr. Henry Earle, of London, reported, at a much earlier period than Dupuytren, in the "Medico-Chirurgical Transactions," vols. v. and vii., cases of this injury, involving the neck and upper limbs, successfully treated by the same methods. Mr. G. H. James, Surgeon to Dover and Exeter Hospital, subsequently to Mr. Earle, reported, in vol. xiii. of the "Medico-Chirurgical Transactions," cases successfully treated on the same plan; and in a recent publication "On the Results of Operations for Cicatrices after Burns," London, 1868, Mr. James has given his additional experience.

CASE XXVI.—*Disfigurement of the Face from Cicatricial Contractions; Extirpation of one Eyeball; Closure of the Orbit by a Plastic Operation.*

ANNIE MULLADY, aged ten, of Irish parentage, from Scranton, Pa., was admitted, June 2, 1870, into St. Luke's Hospital. The burns, from the consequences of which her present disfigurement resulted, occurred four years previously in a railroad-

car that was precipitated over an embankment, causing a heated stove to fall upon her. Her condition
at the time of her admission was as follows: Both
lids of the left eye were entirely destroyed, leaving
the eyeball uncovered and projecting beyond the
orbit. The eye itself was sightless from dense opaci-
ty of the cornea. A condition of extreme irritability
kept it in constant motion and affected the sound
eye, obliging her habitually to incline the head forward, so as to shun direct exposure to the light.
An ulcerated spot, of the size of a silver half-dollar
piece, still remained unhealed over the left frontal
boss. The entire surface of the left half of the forehead, surrounding the ulcer, was cicatricial, and the
same condition extended up over the scalp, which
was denuded of hair-growth. The surface of the
left cheek and temple, as well as the nose, was in
a like cicatricial condition. The nose itself was
shrunken, distorted, and drawn over toward the left
side to such a degree that the columna nasi no longer
corresponded to the inferior border of the septum,
but stood off to the left of it, leaving it exposed to
view. Both nasal orifices were misshapen and contracted. The upper-lip border, at the left angle of
the mouth, was drawn up into a notch. The inner
half of the upper lid of the right eye was everted
and notched by a cicatricial contraction of the skin
above it. The right eye itself had escaped injury,

and possessed good vision. (*See* Fig. 74.) Her general health was good. The first and most urgent indication was to relieve the sound eye of its sym-

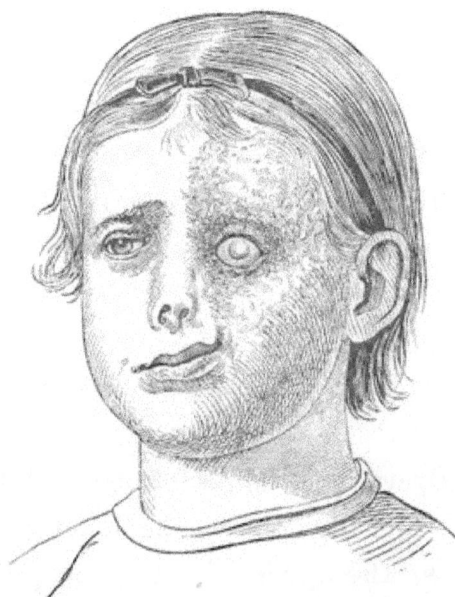

Fig. 74.

pathetically-induced irritable condition, by the extirpation of the sightless ball of the left eye.

First Operation was performed June 29th, as follows: The remains of the conjunctiva were detached from the ball of the eye with scissors, and the muscles successively divided at their insertions into the globe of the eye. The optic nerve was then severed at its junction with the eyeball, and the ball itself removed. No arteries required to be ligated.

A covering of soft scraped lint was laid over the part, and secured by a single turn of bandage, lightly applied around the head. Healing progressed gradually, and a slightly-concave surface, bounded by the bony margin of the orbit, remained; the surface consisting, as it did, of conjunctiva, continued to exude a viscid secretion. The effect upon the sound eye of the removal of its fellow was most beneficial. It at once regained a quiescent condition, and was no longer sensitive on exposure to the bright light. The next step to be undertaken was to restore the upper lid of the right eye, which was everted at the inner half of its tarsal margin. For this purpose a

Second Operation was performed, as follows: An incision, commencing above the inner canthus, was carried across the upper lid on a line parallel with the commissure of the lid, to a point upon the right temple above and a little beyond the outer canthus. The everted inner half of the tarsal edge of the upper lid was liberated, and brought down into contact with the opposite tarsal margin of the lower lid. and the edges of the wound were dissected up to afford space for the insertion of new material, which was obtained by raising from the temple a patch of skin, having its pedicle of attachment adjacent to the outer extremity of the bare space that was to be covered, and its long axis extending upward on the temple. The outlined patch, after being dissected up from its underlying connections, was transferred edgewise (by causing it to perform a circuit of a quadrant) to the space above the upper eyelid, and there accurately adjusted by sutures. The raw surface left upon the temple was covered by approximating the opposite edges of the wound, and securing them in contact also by sutures. Primary union followed, and the notch at the tarsal margin was nearly obliterated. The conjunctival surface covering the left orbit, from which the eyeball had been extirpated, continued to exude a viscid secretion, and presented a conspicuous, unsightly disfigurement. It was therefore decided to conceal it by covering it with a patch of skin, to be taken from the right half of the forehead—that being the nearest point where sound material could be obtained for the purpose. As a preparation for the proposed operation, the secreting conjunctival surface covering the orbit required to be converted into a granulating surface. This was done by the application of a cautery-iron, heated to redness, and was followed by the separation of a superficial eschar, and the substitution of a healthy suppurating surface. After this preparatory step a

Third Operation was undertaken December 22d, as follows: An incision, dividing the skin, was carried around the bony margin of the left orbit, and the outer edge of the incision was dissected up and everted sufficiently to allow it to be adjusted to the edge of the patch of skin with which the included space was to be covered. To provide material for this purpose, a patch of skin of the requisite size and shape, regulated by a pattern previously cut from oiled-silk, was outlined upon the right half of the forehead by an incision, the long axis of the patch being directed upward, and its pedicle of attachment corresponding below to the inner half of the eyebrow. Its free extremity included a portion of the hairy scalp above. In order to afford a continuous raw surface between the pedicle of the patch and the bare orbitar space it was intended to cover, an incision was carried from the edge of the patch at the inner end of the right eyebrow transversely across the forehead to the bare orbitar space, and the skin intervening between the patch and space was displaced from above the incision toward the left side, where it retained its attachment, and was reserved for subsequent use. The patch was now brought around edgewise, from right to left, and from above downward, till it reached its destination, and was adjusted by its under surface continuously in contact with the underlying denuded surface prepared for it. The adjustment was secured by pin and thread sutures. The displaced portion of skin, held in reserve, was utilized by transferring it in an opposite direction, from below upward, and from left to right, to the surface left bare by the forehead-patch, the lower half of which it filled up, leaving the upper half to be treated with a collodion crust, under which healing might take place by granulation.

The subsequent management and progress require no special notice. Primary union took place at almost all points of the circumference of the patch. A slight discharge of pus, mixed with lachrymal secretion, continued for several weeks to escape from the orbit by a small opening at its outer margin, but at length it dried up spontaneously, and the opening itself finally closed permanently. The collodion-crust became detached and was removed on the tenth day. The underlying surface was in a healthy granulating condition, and thereafter cicatrized progressively till it finally healed. Attempts were next to be made to remedy the remaining deformity of the face, especially the distortion of the nose and upper lip, at the left angle of the mouth. On December 24, 1870, a

Fourth Operation was performed, as follows: An incision, commencing at a point one finger's-breadth below the inner canthus of the right eye, was carried obliquely downward across the nose and left cheek to a point one inch below the left angle of the jaw. In its entire course this incision divided the cicatricial surface. Both edges of the incision, after being dissected up, receded wide apart, so that the nostrils and the left angle of the mouth regained their natural shape. In order to fill up the wound-space thus

13

produced with new material, a patch of skin of the
same shape and size was outlined by an incision upon
the left cheek and temple, and dissected up from its
underlying connections, but left attached by a broad
pedicle below, where it was adjacent to the extremity
of the wound-surface which it was intended to cover.
The patch itself was now transferred edgewise to its
new locality, and the confronted edges of the patch
and space were accurately adjusted to each other and
secured by sutures. Omitting unessential particulars,
it is sufficient to state that the transferred patch
sloughed, owing no doubt to the cicatricial condition
of its surface. After the sloughs had separated and
the surface assumed a healthy granulating condition,
another attempt was made to supply new material
with which to fill the wound-space. This was attempted by engrafting a patch of skin from the patient's own hand. For this purpose a

Fifth Operation was performed April 14, 1871,
as follows: The wound-surface upon the left side of
the nose and cheek was first prepared by paring away
the granulations with scissors applied flatwise, and
trimming the edges afresh after dissecting them up.
A patch of skin of the required shape and size was
then outlined upon the dorsum of the patient's own
left hand, and located with its pedicle resting upon
the dorsal surface of the metacarpal bone of the
thumb, and its free extremity at the commissure be-

tween the index and middle fingers. The patch was dissected up from its underlying connections, and left attached at its pedicle. The wound-surface on the back of the hand, made bare by raising the patch, was closed by approximating the opposite edges of the wound, and securing them in contact by sutures, before putting the hand in position. The hand itself was now brought up with its palmar surface applied to the left side of the face, and supported in position, while the patch from the hand was transferred to the cheek, and accurately adjusted to the space prepared for it, and there secured by sutures. A succession of assistants was organized to support the hand and arm in position, and maintain them at perfect rest, if possible. At the expiration of forty-eight hours, however, patient could no longer endure the irksomeness of her position, and was relieved by severing the patch at its pedicle. Its changed color had already indicated that its vitality had ceased, and the experiment had failed. The wound upon the dorsum of the hand healed mostly by primary adhesion, and no defect in the functions of the fingers ensued. Not willing to abandon my purpose, another operation was devised, with the view of obtaining sound skin from below the jaw, with which to cover the wound-space upon the cheek. On the 6th of May following a

Sixth **Operation** was executed, as follows: The wound-surface on the left cheek having again assumed a healthy granulating condition, it was prepared in the same manner as for the preceding operation. A patch of skin of the requisite shape and size was then taken from the side of the neck under the jaw, the free extremity of the patch extending forward beyond the symphysis under the chin, and its pedicle retaining its connection over the angle of the jaw. After the transfer of the patch to its new locality upon the cheek, and its adjustment by sutures, the wound-surface below the jaw was closed by approximating the opposite edges of the skin, and securing them in contact by sutures. About three-quarters of an inch of the extremity of the patch sloughed within forty-eight hours after its transfer, and primary union failed to take place elsewhere. It therefore became necessary to maintain the remainder of the patch in place during the process of healing by granulation, and especially to hold it well up toward the nose. To accomplish this a beaded-wire clamp-suture was employed in the following manner: A common darning-needle, threaded with flexible silver wire, was passed through the middle of the patch of skin, and then carried through the outer wall of the left nostril, close to the cheek, and into its cavity and out again by its external orifice. A smooth round glass bead was strung upon the end of the wire coming out of the nostril, and the end of the wire itself knotted to hold the bead. The other end of the wire traversing the patch of skin was also strung with a bead and a perforated leaden shot. By drawing upon this end of the wire, the bead at the opposite end was brought into the cavity of the left nostril, and rested in contact with its outer wall. At the same time the patch of skin was slid up against the side of the nose and cheek, then secured in its place by the bead pressed against it, and held fast by mashing the shot upon the wire with pliers. This contrivance answered a very useful purpose in keeping the patch quietly in place during the process of healing by granulation. The adjustment of the parts was further perfected by compresses kept in place by strips of adhesive plaster, often renewed. The beaded-wire suture was left in place till the twelfth day, when healing was so far advanced that it could be dispensed with. A plug of soft sponge of suitable size was kept in the left nostril to maintain it in good shape, and was renewed daily. On June 9th cicatrization was complete. On the 15th her mother took her home without my knowledge or consent. This abrupt removal from the hospital cut short all further treatment by which a more perfect result might have been attained. On May 19, 1872, I saw patient at her home, in Scranton, and ascertained as follows: There were scarcely any remains of the notch in the upper lip at the left angle of the mouth. The nose was less distorted than before the treatment. The patch covering the left orbit, consisting as it did of a portion of the hairy scalp, was covered with a growth of hair. This it was intended to destroy had she remained longer in the hospital.

Fig. 75.

Fig. 75, showing the final result, was from a photograph taken at a gallery in Scranton, at my request.

CASE XXVII.— *Cicatricial Contractions involving the Chin and Front of the Neck.*

ALEXANDER S., residing with his parents in the upper part of this city, was, in February, 1869, at the age of two years and one month, severely burned from his clothes taking fire. The chin and front of the neck were the parts most seriously involved. In the month of June following, the burnt surface having healed, patient accompanied his family to their residence in the country. In October, 1870, when he came under the author's care for surgical treatment, being at the time three years and nine months old, his condition was as follows: A broad band of cicatricial skin of dense structure occupied the front of the neck, extending in a direct line downward, from the lower border of the under jaw to the top of the sternum and clavicles, and approximating the parts so that the jaw could not be elevated beyond four inches above the sternum. The connections of the band with the jaw laterally reached from the left angle of the jaw to a point below the right angle of the mouth. Its connections below with the top of the sternum and upper surface of the clavicles corresponded in extent laterally with those above. The prominence of the chin and the profile outline of the front of the neck were both obliterated. The separation of the jaws and lips from each other left

the tongue habitually exposed to view, and allowed
the saliva to dribble constantly from the mouth.
The anterior surface of the band was uneven, and
traversed in a vertical direction by alternate ridges
and furrows. From both edges of the band sound
skin extended upon the sides of the neck, after first
receding behind the band and forming there a pocket
of such a depth that the fingers, when pressed toward
each other from opposite directions behind the band,
could be made to meet together in the median line,
with the skin only intervening between them. When
the band was put upon the stretch by elevating the
head, the left half of the under lip was drawn down
more than the right. Patient also suffered from a
tormenting itching, that would suddenly attack the
cicatricial parts, and provoke him to scratch and pull
upon the band in an almost frantic manner. His
mother was of the opinion that for several months
past the condition of the parts had undergone no
change. His general health was good. (*See* Fig.
76.)

First Operation. — Performed at patient's residence, November 5, 1870, with the aid of Prof. A. C.
Post and Drs. C. M. Bell, J. N. Beekman, and Robert
Watts, as follows: The entire cicatricial band was
divided into three serrated angular flaps by two diverging incisions, carried from the symphysis menti
downward and outward to either lateral margin of

the band, where it joined the clavicles.   From these
terminal points incisions were made, one along either
margin of the band, upward and outward to the
lower edge of the jaw.   The three flaps thus formed
were then dissected up from the subjacent loose connective tissue, beginning at their apices, and proceeding toward their bases, at which latter points
the dissection was carried a short distance beyond

Fig. 76.

the limits of the cicatricial skin.   The head could
now be thrown back and rotated freely in every direction.   The surface thus laid bare involved the
entire anterior region of the neck.   The next step
was to reapply and adapt the detached flaps to this
extensive denuded surface, while the head was kept
in an elevated position.   In effecting this adjustment
it was necessary to excise redundant folds and pare
off the edges of the flaps, in order to adapt them to

each other. On the left side of the neck it was also found necessary, in order to relieve tension and facilitate the approximation of the edges of the flaps, to make an incision, five inches long, through the thickness of the skin across the base of the neck, on a line parallel with the edge of the wound and two inches distant from it. The edges of this incision gaped widely apart, and thereby the desired end was accomplished. The raw surface, resulting from the side incision, was coated over with collodion-crust and left to heal by granulation. A covering of scraped lint was spread over the other parts, and secured by strips of adhesive-plaster. Though the operation was necessarily protracted, and the loss of blood considerable, there was no extreme depression of the vital powers in consequence. A good degree of reaction followed, and, aided by an anodyne, the patient passed the first night pretty comfortably. At the expiration of forty-eight hours sloughing had destroyed all the flaps, except a portion about one inch in breadth along their bases. The sutures were removed in succession, and the sloughs got rid of as fast as they separated. Special care was taken, in dressing the wound, to hold the detached skin, that had escaped sloughing, in close contact with the subjacent surface by means of long strips of adhesive-plaster, carried high up on either side of the face and over the temples, and on the sides of the neck. At the same time the head was also kept well elevated. The wound took on healthy action after the separation of the sloughs, and the patient's general condition was all that could be desired. The detached skin became adherent, and cicatrization progressed favorably. The exuberant growth of granulations was repressed by the energetic use of solid nitrate of silver, not merely passed over the surface, but plunged deep into the substance of the granulations. This was sometimes repeated daily. Solid caustic potassa was also applied, but at much longer intervals, and only at points where the growth was not sufficiently controlled by the nitrate of silver. At the expiration of about four weeks, when the dimensions of the sore had considerably diminished, a stiff leather stock, protected by a covering of canton flannel, was adapted to the neck and worn constantly, so as to keep the head elevated and oppose the disposition to recontraction in the direction of flexion. As cicatrization advanced, the newly-formed cicatricial tissue manifested a tendency to form salient corded bands, which, if left uncontrolled, would have reproduced to a greater or less degree the original deformity. To prevent this effect, the bands were divided at two or more points across their entire thickness, and to a short distance on either side through the neighboring skin, and deep enough to expose the subjacent loose connective tissue, thus

permitting the fresh-cut edges to retract widely apart.
This had the effect of breaking up the continuity of
the bands and neutralizing their action. These operations were repeated successively, with the aid of

Fig. 77.

etherization, on the 7th, 12th, and 29th of December.
The leather stock, worn for the support of the head,
proved after a time objectionable, on account of its
chafing the skin and producing ulceration. While

endeavoring to devise some substitute for the stock,
my attention was directed to a brace, used by Drs.
W. E. Vermilye and C. T. Poore, in the treatment of
caries of the cervical vertebræ, which seemed admirably suited to my purpose. At my request, Dr. Poore
adapted one to my patient. On the 21st of January
it was applied, and has been worn constantly since,
except at night. It consists (*see* Fig. 77) of two
padded steel bands, arranged parallel to each other,

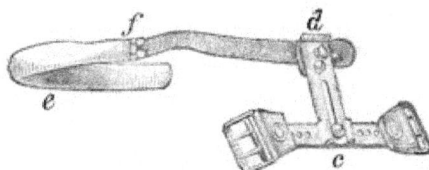

FIG. 78.

one on either side of the spine, and adapted flatwise
to the natural curve of the back. It extends lengthwise from the last cervical vertebra to the top of the
sacrum.

These vertical bands are joined below by a broad
padded metallic band, which passes half round the
body behind and just above the hips. At their upper
ends the two vertical bands are joined by a cross-piece,
c (Fig. 78), to which a steel ring or collar, d, of an
oval shape, is joined by an upright piece, $c\,d$, in such a
manner as to stand horizontally, and afford a support
in front to the chin. A segment of the ring in front,
e, where it corresponds to the chin, is covered with

chamois-leather, and forms a shelf for the chin to rest upon. On one side, near its middle, the ring has a hinge-joint, f, which permits it to be opened in two halves, and hence facilitates its removal and re-application. By means of a screw at the joint over

Fig. 79.

the nape of the neck, a lever action is made to elevate the ring in front and regulate the height of the chin. Two shoulder-straps, gg (Fig. 77), and an apron, h, with three straps at either lateral edge, serve to fasten the brace in close contact with the body.

The brace, besides supporting the chin in an elevated position, and thereby resisting the recontraction of the cicatricial formation in a vertical direction, exerts, by means of the straps which pass over the front of the shoulders, a constant outward traction upon the skin covering the lower part of the neck and upper part of the chest, the effect of which is also to resist contraction laterally and keep the cicatricial surface flat and smooth. Another important advantage of the brace is, that it compels the patient, whenever he wishes to move his head in any direction, to elevate it so as to clear the chinpiece, or, in other words, by voluntary muscular action to stretch the cicatricial surface in front. In addition to the brace, patient has worn constantly, night and day, a cravat of canton flannel, two fingers wide, secured around and in close contact with the neck by an elastic strap and buckle, for the purpose of holding the new cicatricial surface in contact with the subjacent parts. From its first application the patient has worn the brace uninterruptedly, except at night, and so comfortably as scarcely to restrict his activity or enjoyment. A progressive improvement has taken place in the under lip. It has regained its natural form and position, and the saliva no longer escapes uncontrolled.

Fig. 79, copied from a photograph taken April 22d, shows the final permanent result. The contour of the chin and front of the neck were restored to their natural form and dimensions, and the head enjoyed entire freedom of motion in all directions. The newly-healed surface presented cicatricial lines radiating in different directions, downward as low as the second rib, upward as high as the left angle of the jaw, and also on both sides of the neck. The cicatricial parts were mostly smooth and level, and pliable upon their underlying surfaces. During the hot weather of August, 1871, the brace was left off entirely for a time, and without any detriment. Subsequently, by way of precaution, it was resumed and worn two days in the week for a while longer, and then dispensed with altogether. On January 1, 1874, patient was examined, and a further improvement found to have taken place. The skin covering the front of the neck and chin had become more supple and pliable. The under lip still maintained its normal relations to the upper lip, and performed all its functions, and the head enjoyed the utmost freedom of motion in all directions.

CASE XXVIII. — *Cicatricial Contractions involving the Face and Hand.*

H. C. W., aged four years and seven months, a resident of New Jersey, of sound constitution and enjoying good health, was extensively burned, when

sixteen months old, by overturning upon himself a
kerosene study-lamp. At the expiration of about
fourteen months the burnt surfaces had mostly healed.
The parts involved in the cicatricial contractions
which resulted from the burns were, the left half of
the face and scalp and the dorsal surface of the left
hand and forearm. When I first examined him, in
February, 1872, his condition was as follows: The
left half of the scalp, as far back as the occipital
region, was bare, and the surface presented a pale,
shining, cicatricial aspect; the skin, however, was
pliable, and moved freely on the underlying parts.
The left ear, though diminished in size by the loss
of its rim, retained its natural shape, and was not
adherent to the scalp. The skin covering the left
half of the forehead and temple, though cicatricial
on its surface, was still movable on the subjacent
parts. The outer half of the left upper eyelid was
everted to an extreme degree, and spread out upon
the eyebrow; the inner half of its tarsal border alone
came into contact with the lower lid when they were
closed. The conjunctiva covering the everted portion
of the lid being relaxed and swollen, filled up the
space between the lids when closed, and thus protected the cornea, which otherwise would have become opaque from constant exposure. The eyelashes,
as well as the tarsal edges of both lids, had escaped
injury. The eyebrow was denuded of hair. The
14

eyeball itself had sustained no damage, the cornea retained its natural lustre, and vision was unimpaired. The surface of the left cheek and side of the nose was cicatricial, and the contraction consequent thereupon had drawn the under lid away from contact with the eyeball, but without producing any eversion of its tarsal margin. The left angle of the mouth was somewhat drawn upward by a vertical fold of cicatricial skin upon the cheek, immediately above it. In consequence of the condition of the lids of the left eye, patient habitually held his head inclined forward and to the left side, so as to avoid direct exposure to bright light, and had thereby acquired a peculiar expression of countenance (*see* Fig. 80). A description of the left hand and forearm will be given in a subsequent part of the narrative.

On the 6th of February, 1872, an operation was performed in the presence of Dr. James A. Davis, the family physician, from Bloomfield, N. J., Drs. John N. Beekman and Thomas E. Satterthwaite, and Prof. A. C. Post, M. D.

First Operation.—The object of this operation was, to restore the upper lid of the left eye to its normal relations and functions. It was attempted as follows: Two incisions were started from a single point high up on the forehead, above the middle of the left eyebrow, and continued downward in lines diverging from each other and terminating, one at either canthus of the eye. The inverted V-shaped patch of skin, included between these incisions, was dissected up from the pericranium as low down as the bony margin of the orbit. The upper lid was thus liberated and brought down so as to allow its tarsal margin to be adjusted in contact with that of the under lid. Before proceeding further, a trans-

FIG. 80.

verse fold of the redundant conjunctiva, lining the upper lid, was excised as far back from its tarsal margin as possible. With the descent of the upper lid, the inverted V-shaped patch of skin, which had been raised from the forehead, was brought down and secured at a lower level by a pin suture, inserted on either side, together with additional thread sutures above the pin sutures. The surface left bare on the forehead by the descent of the patch was closed by approximating the opposite edges of the wound, and securing them in contact by thread sutures. To relieve these last sutures from tension, parallel incisions were made through the skin on either side, the one over the left temple being three inches, while the other, toward the middle of the forehead, was only one inch in length. The edges of these incisions gaped widely apart, and the desired result was thus obtained. The bare surfaces left were coated with a collodion-crust. In order to hold the tarsal edges of the lids more exactly in coaptation, a beaded silver-wire clamp-suture (*see* page 17) was passed in a vertical direction through both lids, near the outer canthus, and out of the way of the cornea, and left *in situ* for three days. Wet dressings were avoided, and a layer of woven lint of double thickness was kept applied to the parts. Moderate febrile reaction followed the operation, but it subsided on the third day. On the fourth day the last sutures were removed. Sloughing, however, had already taken place, and had involved about three-fifths of the transplanted patch. Healthy suppuration succeeded the separation of the sloughs, and the ulcerated surface progressively diminished in size. The collodion-crust separated from the forehead on the sixth day, and from the left temple on the tenth, leaving healthy granulating surfaces to heal by cicatrization. The eyeball had in no way suffered from the presence of the silver wire which traversed the lids. Patient rapidly regained his spirits, and, at the end of one week, resumed his accustomed amusements, enjoyed his meals, and rested well at night. The sloughing which had taken place was no doubt to be attributed to the cicatricial condition of the transplanted patch of skin. The ulcerated surface above the left eyebrow progressively diminished in size, and, at the end of the fourth week, measured one inch in its transverse diameter, by three-fourths of an inch vertically. If left to itself it was feared that the contraction consequent upon cicatrization might reproduce, to a greater or less degree, the original eversion of the upper lid. To prevent this it was determined to transfer a portion of sound skin from the right half of the forehead and engraft it upon the ulcerated surface. The operation for accomplishing this object was performed on the 8th of March.

Second Operation.—The ulcerated spot itself was prepared by first excising the granulating surface down to the level of the pericranium, with scissors, applied flatwise, and then paring afresh the edges and everting them slightly. A transverse incision was then carried across the forehead, on a line continuous with the lower margin of the spot just prepared, and as far as the inner extremity of the right eyebrow. The upper edge of this incision was dissected up to afford a bare surface of one finger's-breadth, which would be continuous with and form a part of the space above the eyebrow. A pattern, cut from oiled silk, of the size and shape of the space just prepared, was applied upon the right half of the forehead, in a vertical position, with its base resting upon the inner half of the eyebrow, and its free extremity involving the hairy scalp above, which had been previously shaved clean. An incision was then carried around the margin of the pattern, and the included underlying patch of skin was dissected up from the pericranium, but left connected, for support, at the margin of the orbit. The pedicle of the patch, at its inner edge, toward the median line, was adjacent to the bare surface which it was intended to cover. Additional room had to be made for the transfer of the patch, by dissecting up the skin from the forehead, above the nose, and displacing it toward the left side, where it remained attached, and was reserved for subsequent use. The patch was now brought down edgewise from right to left, and adjusted accurately to the edges of the space prepared for it, with sutures inserted close together. In order to utilize the portion of skin displaced from above the nose, for the purpose of covering the surface left bare on the right half of the forehead, it was carried upward from left to right and adjusted to the lower part of the bare surface by means of sutures. The remaining upper portion of the bare surface was coated over with a collodion-crust, and left to heal by granulation. Before proceeding to the operation just described, a transverse fold was excised, for the second time, from the still redundant conjunctival lining of the upper lid, and the lids themselves were secured together by a single-thread suture inserted through the skin alone, near their tarsal edges, and toward the outer canthus. A strip of woven lint, saturated with collodion, was applied across the outer half of the closed lids, to afford additional support. The entire operation occupied one hour and a half, and was well borne, the loss of blood having been inconsiderable. A layer of woven lint, of double thickness, spread with cerate to prevent its sticking to the surface, was laid upon the forehead. Elixir opii, gutt. x, and weak brandy-and-water were prescribed.

March 10th.—*Second Day.*—Slight febrile reaction, and moderate inflammatory tumefaction. Removed three pins, and changed the yarn on the remaining ones.

12th.—*Fourth Day.*—All the sutures have been removed in succession, including the one holding the

lids in contact. To supply the place of this last
suture strips of adhesive-plaster were applied. Primary union has taken place at almost all points, and
without any sloughing.

20th.—Twelfth Day.—The collodion-crust came
off from the forehead, exposing healthy granulations
at the circumference of the sore, but in the centre a
brownish patch of sloughing pericranium, which had
been unintentionally divided in the operation. No
exfoliation of bone, however, followed the sloughing
of the pericranium ; healthy granulations covered
the spot, and cicatrization followed. Iron and qui-
nine were ordered as a tonic.

Third Operation.—The patient being in excellent
condition, from a stay of three weeks in the country, a third operation was performed on the 20th of
April, for the purpose of removing a conspicuous
distortion of the left angle of the mouth. An incision, commencing at a point on the left cheek, bordering on the middle of the nose, was carried downward and outward across the cheek to a point a little
anterior to the angle of the jaw. In its course the
incision divided the cicatricial fold which drew up
the angle of the mouth, and so allowed the latter to
regain its natural shape. The edges of the incision,
after having been dissected up, receded from each
other and left a space between them of about one
finger's-breadth. To fill this up with sound skin,

the following method was adopted: A patch of skin
of the required shape and size, with its pedicle of
attachment adjoining the space to be filled up, and
its free extremity on a line below the symphysis, was
dissected up from the side of the neck under the jaw.
This patch was then transferred edgewise to the
space prepared for it, and there accurately adjusted
by sutures. The wound left on the neck was closed
by approximating its opposite edges, and securing
them together with sutures. The treatment was the
same as after the previous operations. Union failed
to take place, and sloughing of about three-fifths of
the patch followed.

On the fourth day the slough had separated, and
healthy suppuration succeeded. It was now important to prevent any shrinking of what remained of
the patch, and to maintain it in place. This was
done by adhesive straps carefully adapted and frequently renewed.

April 25*th.*—A mild attack of erysipelas developed itself upon the left ear and neighboring surface of the scalp, but soon passed off without any
serious consequences. From this time his general
health improved, and cicatrization of all the sore
surfaces progressed steadily till June 8th, when all
had finally healed. The result of the last operation,
notwithstanding the loss of so large a portion of the
transplanted patch of skin, was a certain degree of

improvement of the angle of the mouth. As no further operations could well be undertaken till the au-
tumn, patient's nurse was instructed to manipulate
the parts upon the forehead and left cheek daily, so
as to increase their pliability and prevent contraction
from taking place. The good effect of these manipulations was manifest when patient returned in October to spend the winter in the city with his family.

The distortion of the left hand, the treatment of
which was now to be undertaken, was the result of
the same burn that had disfigured the left half of
the face. The condition of the hand was as follows:
The thumb and all the fingers, though flexed at their
phalangeal articulations, were drawn backward toward the dorsum of the hand to such a degree that
the proximal ends of the first phalangeal bones rested
on the dorsal surfaces of their supporting metacarpal
bones. This distorted condition, which existed to a
greater degree in the thumb and index than in the
other fingers, was maintained by salient folds, consisting of cicatricial skin alone, that were given off
from the dorsum of the wrist in radiating lines, one
fold proceeding to the thumb and each finger. While
the middle, ring, and little fingers retained their parallel position toward each other, the index was widely
abducted from the middle finger, and the thumb
from the index-finger.

The hand, notwithstanding its distorted condition,

still performed useful service. Flexion and extension
at the elbow-joint, as well as pronation and supi-
nation of the forearm, remained unimpaired. (*See*
Fig. 81.)

FIG. 81.

Preparatory to an operation, a guttered splint,
made of tin, was adapted to the palmar surface of
the forearm and lengthened out at the wrist by the
addition of a flat piece, bent flatwise in the direction
of flexion, and adapted for the support of the hand.
On the 24th of October, 1872, a fourth operation
was performed.

Fourth Operation.—The second, third, and fourth

fingers being similarly involved, their treatment only
was to be attempted in the present operation, which
was executed as follows: By flexing the fingers and
putting on the stretch the cicatricial folds on the
dorsum of the hand, a broad longitudinal fold of
skin was also rendered tense and salient upon the
forearm above. This fold was transfixed at its base,
near the middle of the forearm, while gathered up
between the thumb and fingers, and a tongue-shaped
flap of skin was formed by cutting downward toward
the wrist and outward through the surface. The detached flap receded toward the elbow, and the three
fingers could now be flexed to a right angle at their
metacarpo-phalangeal articulations; the phalanges
themselves, however, could not be flexed upon each
other, owing to the remaining resistance of the edges
of the wound just made upon the forearm, which
became very tense and unyielding whenever flexion
was attempted. These edges were therefore freely
divided across at selected points till there remained
no longer any resistance, and complete flexion of all
the phalanges could be effected. No tendons or
muscles were exposed; the surface laid bare consisted of adipose and connective tissue. A single
vessel only required to be ligated. Here and there
a suture was inserted to hold the detached skin in
place, but without any attempt to procure adhesion.
The limb was then placed upon the splint prepared

for it, and the three fingers were secured in a flexed position, by adhesive-plaster, to the flat piece that joined the splint at an angle of flexion, opposite to the wrist-joint. The raw surface, which involved the dorsum of the hand and lower two-thirds of the forearm, was coated with collodion-crust.

After completing this operation, advantage was taken of the anæsthetic condition of the patient to perform another operation for the further improvement of the lids of the left eye. The outer half of the tarsal edge of the upper lid was still disposed to become everted, and this tendency was promoted by the presence of a mass of granulations in the conjunctival *cul-de-sac* at the outer canthus. The commissure of the lids had also become lengthened in a marked degree, so as to exceed the length of the commissure of the lids of the other eye by about one-quarter of an inch. The mass of granulations was first excised by seizing it with a fine-clawed forceps and clipping it off at its base with a scissors curved flatwise. The attempt was next made to bring about permanent adhesion between the tarsal edges of the lids at the outer canthus. While the lids were held wide apart, and the eye itself was protected by a wet rag, a ball-shaped cautery-iron of the size of a pea, heated to redness, was buried in the *cul-de-sac* of the conjunctival cavity, at the outer canthus, and applied thoroughly to both tarsal edges of the lids for a distance of nearly two lines from their junction with each other. Compresses wet with ice-water were immediately applied, and afterward frequently renewed.

On coming out from the anæsthesia, patient did not appear to suffer from this severe application. No febrile reaction followed. He passed a good night without an anodyne, and the next day sat up with his arm supported in a sling. Scarcely any redness or swelling of the hand or forehead supervened.

On the fifth day, October 29th, the collodion-crust came off from the wrist and forearm, leaving a healthy granulating surface, which thereafter was dressed with simple cerate, the limb being kept upon the splint. A moderate degree of inflammation followed the application of the actual cautery to the conjunctiva and tarsus. No injury was sustained by the cornea or ball of the eye. Superficial eschars separated, and healthy suppuration followed. Upon the forearm portions of cicatricial skin, which had been detached from their underlying connections, sloughed.

November 5th.—The exuberant granulation-growth has required the energetic application of solid nitrate of silver at every daily dressing. Forcible flexion of the fingers and wrist has been daily practised: this, though a severe process during its performance, was not followed by any lasting pain afterward.

It now became apparent that a great advantage would have been gained by liberating the index-finger, and restoring it to its normal relations at the same time with the other three fingers; it was accordingly decided to do it without further delay. The operation was performed on the 5th of November.

Fifth Operation.—Two incisions, beginning one at the commissure between the thumb and index-finger, the other between the index and middle fingers, were carried through the skin, upward, in converging lines, until they met at a point above the cicatricial fold of skin which held the finger extended. After these incisions had been made, the index-finger could be flexed at its metacarpo-phalangeal articulation, but not at its phalangeal joints. By dissecting up the skin on the dorsal surface of the first phalanx from its underlying connections, the power of flexing the finger at these articulations was fully restored. The index-finger was then brought down by the side of its fellow, where it was secured, and thereafter treated in common with them.

The cauterized surfaces at the outer canthus of the left eye being now in a state of healthy suppuration, and all swelling having subsided, the tarsal edges of the lids were secured in exact contact by a beaded silver-wire suture (*see* page 17) inserted in

a vertical direction through both lids, at a distance
of half an inch from their tarsal edges, and at the
same distance from the outer canthus.   Two fine-
thread sutures were also inserted at the edges, after
first scraping them with a dull-edged knife.   The
thread sutures were removed on the third, and the

Fig. 82.

wire suture on the eighth day.   Permanent adhesion was thus secured between the tarsal edges for a
distance of nearly three lines from the outer canthus, thus shortening the commissure so as to correspond with that of the other eye.

Early in the month of January, 1873, the wound-surfaces on the dorsum of the hand and wrist had healed. The manipulation of the wrist and finger joints had been continued daily with decided benefit, and the limb had been kept constantly secured to the splint. The thumb, not having yet been subjected to any treatment, remained unchanged. To remedy this deformity was the object of the

Sixth Operation, *February* 12, 1873.—By stretching the thumb in the direction of flexion, a longitudinal fold of cicatricial skin was brought into prominence upon the radial border of the wrist and forearm. At first broad and embracing the root of the thumb, this fold grew narrow as it extended obliquely upward on the palmar surface of the forearm, in the middle of which it became corded, and continued so to the elbow. This fold in the process of its formation had evidently drawn up the thumb into its distorted position, and was still the chief obstacle in the way of its performing complete flexion. This fold was divided, while held on the stretch, by a transverse incision passing one-third around the wrist, and at a distance of one inch above the wrist-joint. The tissue creaked under the knife, and the edges of the wound gaped widely apart, but without affording much relief to the thumb. The subjacent aponeurotic layer was also found tense and resisting, and had to be divided across. Some degree of flexion was thus obtained. It was now as-

certained that the proximal end of the first phalanx
was dislocated, and rested on the dorsal surface of
the end of its supporting metacarpal bone. In order
still further to liberate the thumb, and reduce the
dislocation, a longitudinal incision of the skin was

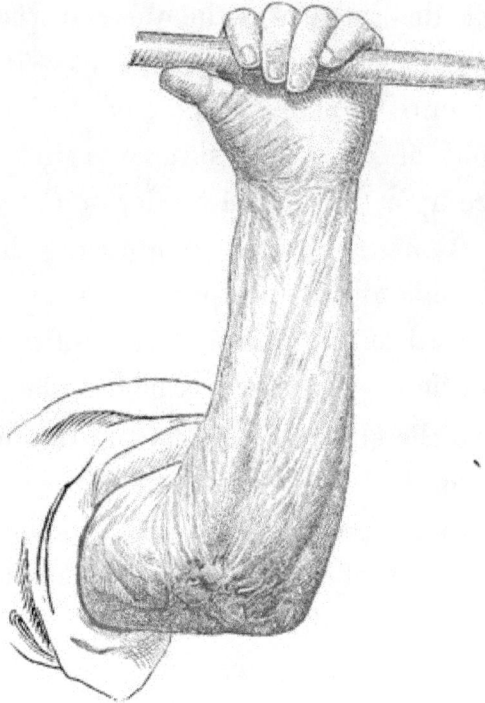

FIG. 83.

carried between the metacarpi of the thumb and the
index-finger upward to join the incision across the
wrist, and the skin was dissected up on the dorsal
surface of the metacarpus of the thumb. All resistance was at length removed, and the thumb could

now be brought down and freely flexed at all its joints. To the splint that had served hitherto for the treatment of the fingers, an addition was made of a separate digitation for the support of the thumb, which required to be maintained in an abducted as well as a flexed position. The newly-made raw surfaces were treated in the usual manner. Patient did well after this, as he had done after the previous operations. The same process of forcible flexion and stretching was employed upon the thumb, as had been used upon the fingers with such good results. In order to reduce the size of the thumb, and improve its shape, it was kept tightly wound with strips of adhesive-plaster. The commissure between the thumb and index-finger was also kept crowded up toward the metacarpus by strips of adhesive-plaster, applied tightly over a saddle-shaped compress made with several thicknesses of woven lint.

On the 1st of May a spot on the wrist, at the root of the thumb, of the size of a finger-nail, was all that still remained unhealed. Under the constant use of the manipulations already described, and the wearing of the splint, the thumb and fingers had regained their natural shape and freedom of motion, and were gradually recovering their power to grasp objects. The thumb, however, in consequence of its greater degree of distortion originally, had not yet recovered the power of flexion so perfectly as the fingers.

The face had also much improved; the commissures of the eyelids were alike in length on both sides, and the tarsal edges of both lids of the left

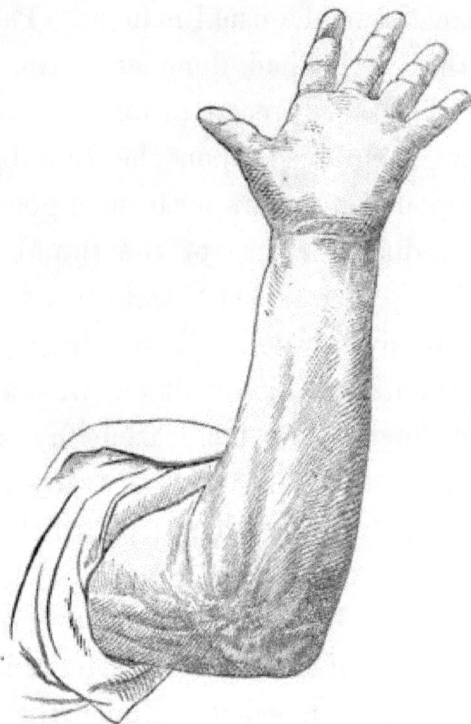

Fig. 84.

eye were perfect in their adjustment to each other. A spot upon the forehead, above the left eyebrow, of the size of a thumb-nail, was covered with a growth of hair that required to be shaved two or three times a week. It was a portion of the hairy scalp, and formed the extremity of the patch of skin which had been transplanted from the right to the left half of the forehead.

On the 10th of May patient accompanied his family on a visit to Europe. From that time to the present (January, 1874), the daily manipulation of the fingers has been kept up regularly, and the splint has been worn at night only, the hand being left free in the daytime to be exercised in every possible way. The father, in a recent letter, says: "His hand has improved, and he now uses the thumb quite a good deal; in fact, for all practical purposes, it is about as useful as the other. We still keep it in the splint at night, and continue the manipulations daily on the hand and also on the face."

Figs. 82, 83, 84, showing the result of treatment, are from photographs taken in Florence, Italy, in January, 1874.

Patient was seen January 15, 1876, and the following particulars ascertained: The lids of the left eye maintain their normal relations to each other, and perform their functions perfectly. A marked improvement of the entire left half of the face is noticeable. The fingers of the left hand perform all their movements to the fullest extent; the thumb only does not admit of flexion to the same degree as the thumb of the other hand, owing to the incomplete reduction of the luxation that had so long previously existed at its metacarpo-phalangeal articulation. This defect does not, however, in any way impair the conjoint use of the fingers and thumb. As a proof of the complete restoration of their functions, he is now taking lessons on the piano, and his teacher regards him as one of his most proficient pupils.

CASE XXIX.—*Cicatricial Contractions involving the Right Axilla and Arm.*

DANIEL CARRIGAN, aged six years, of Irish parentage, was admitted into the Presbyterian Hospital March 17, 1874. ·Four years previously an extensive burn, caused by his clothes taking fire, produced the condition for the relief of which he entered the hospital, and which was as follows: With the right arm elevated to an horizontal position, a fan-shaped fold of sound (non-cicatricial) skin, attenuated to such a degree as to be translucent in the sunlight, was developed between the arm and thorax, with its free border stretching from the elbow to the ninth rib, where it terminated in three small radiating folds. This extensive fold involved and constituted the anterior fold of the axilla. Between it and the posterior fold of the axilla, which did not much exceed its normal dimensions, a deep hollow.

extended high up under the shoulder-joint. The skin covering the anterior surface of the arm and forearm (including the wrist), and of the deltoid re-

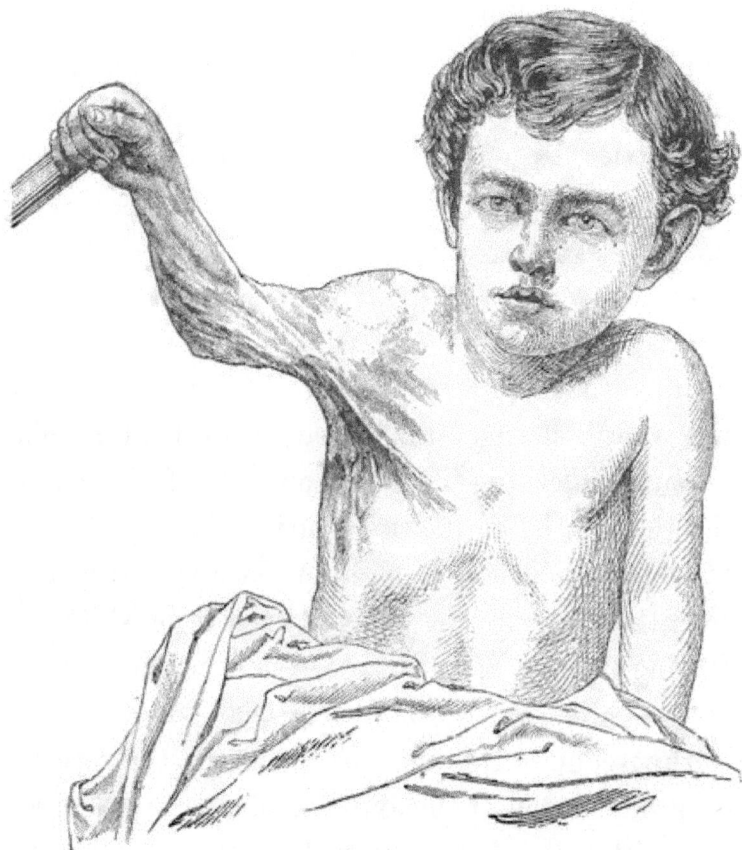

FIG. 85.

gion above, was cicatricial and of unequal thickness, but still pliable and movable on the underlying surface. At the inner margin of the elbow, where the axillary fold terminated, it was continuous with a thickened band of cicatricial skin, which stood out in relief upon the forearm when the limb was put upon the stretch by raising it to an horizontal position. This band extended obliquely down over the forearm to the radial margin of the wrist, and prevented complete extension of the elbow and wrist joints, as well as the further elevation of the arm itself, and it also restricted pronation of the forearm. The shoulder, elbow, and wrist joints, were all free. (*See* Fig. 85.)

First Operation. — This was performed March 26th, as follows: The axillary fold of the skin, being put upon the stretch by elevating the arm, was transfixed with a sharp-pointed straight knife at a point high up under the shoulder, and the fold severed from the arm by an incision continued along the line of their junction, and brought out at a short distance above the elbow. The two thicknesses of skin constituting the fold receded from each other, and spread out over the thorax. The cicatricial band, running lengthwise upon the anterior surface of the forearm as a prolongation of the axillary fold, was next pared away with curved scissors applied flatwise along its base. The arm could now be elevated somewhat higher but not to its fullest extent, owing to the resistance still existing at the edges of the wound just made upon the arm and forearm. These, when further elevation of the arm was attempted, became very tense and unyielding. They were therefore divided across at different points, and also dissected up from their underlying connections until at length all resistance was overcome, and the arm could be elevated to a vertical position, with the elbow and wrist simultaneously straightened out. No muscles or tendons were divided or laid bare; the exposed raw surface consisted of aponeurosis, adipose, and connective tissue. A small islet of cicatricial skin, of the size of a split-pea, was left isolated upon the raw surface at a point near the fold of the elbow, in the hope that it might serve as a point of departure for new growth in the subsequent process of cicatrization. A single small artery only required to be ligated. After hæmorrhage had entirely ceased, the raw surface was covered with a collodion-crust. Patient was then transferred to his bed, and the arm secured in an extended and elevated position by means of a silk handkerchief, rolled diagonally into a cord, which was attached to the wrist and then fastened to one of the crossbars at the head of the bedstead.

27th.—Slept comfortably without an anodyne. Face flushed, moderate febrile heat, and acceleration of the pulse.

28th.—Another good night's sleep; febrile reaction has abated; patient tractable and submissive. Discharge from under the collodion-crust offensive;

removed it entire (on the sixth day), and found the underlying surface of a florid, healthy aspect. Applied simple dressing, with compress and roller-bandage. The arm is kept constantly elevated, and in an extended position.

April 6th.— Commenced dressing the sore surfaces with strips of adhesive-plaster, applied in immediate contact with the surface, the strips overlapping each other like shingles on a roof.

This dressing, which requires to be renewed daily, exerts equable compression over the granulating surface, and represses exuberant growth. It does not, however, suffice alone, but the application of solid nitrate of silver has to be conjoined with it at all points where the growth becomes too luxuriant.

June 10th.—Cicatrization has steadily progressed since the preceding report, and is now nearly completed. On two occasions it has been necessary to resort to the application of solid caustic potassa, the nitrate of silver failing to act with sufficient repressive energy.

A straight, guttered splint, adapted to the dorsal aspect of the entire arm and forearm, has been constantly worn for the purpose of maintaining complete extension at the elbow-joint. In order that the patient might be out of bed, and have the benefit of exercise out-of-doors, a fixture was adapted to maintain the arm constantly in an extended and elevated position. It consisted of a broad belt of webbing, secured around the body by straps and buckles in front. Upon the belt, where it crossed the back, a socket of stiff leather was fastened in an upright position, and, in this socket, a wooden rod was supported erect, with its upper end reaching above the top of the patient's head. The wrist was secured to the upper end of the rod by a silk handkerchief, rolled diagonally into a cord, and the arm by this means was maintained constantly in an elevated position.

July 15*th.* — Cicatrization being now complete, the wearing of the fixtures was discontinued and free use of the arm allowed. Certain gymnastic exercises were, however, to be daily practised, which would tend to stretch and elongate the newly-cicatrized parts in the axilla and about the elbow. Though these exercises were perseveringly practised for five months, and had increased the suppleness and mobility of the newly-cicatrized parts, they had not been sufficient to resist the strong tendency to relapse which characterizes these cases. A certain amount of contraction had again taken place, and two folds of skin, occupying the place of the two natural folds of the axilla, stood out in relief from the surface, and prevented complete extension at the shoulder-joint and also at the elbow. In the hope of remedying this defect, and obtaining a more perfect result, a second operation was performed on the 1st of December.

Second Operation.—The anterior axillary fold of skin being gathered up between the thumb and fingers, and drawn out from the body, its base was transfixed at its lower part with a sharp-pointed, straight bistoury, and an incision carried upward toward the axilla and outward through the surface. A tongue-shaped patch of skin was thus detached, which, left to itself, immediately receded and spread itself out upon the chest, thus permitting the arm again to be elevated to a vertical position. In order to close the newly-made wound, and relieve the sutures that were to unite its opposite edges, an incision four inches long was made parallel with, and three fingers'-breadths distant from, the wound on either side of it. The edges of these incisions gaped widely apart and afforded the desired relief, and the raw surfaces thus exposed were left to heal by granulation after being covered by a collodion-crust. Complete cicatrization of all the parts involved in the second operation was obtained, under the same management as was employed after the first operation. The elevated position of the arm and suitable gymnastic exercises were also perseveringly enforced.

Though a further improvement was obtained by this second operation, there remained still, in the situation of the anterior axillary fold, an elevated fold of skin which was prolonged downward upon the arm, extending over the elbow, and also down on the forearm in the shape of a tense-corded band. When complete extension of the arm was attempted, resistance was made by this corded band. Believing that this remaining resistance could be overcome, I determined on a third operation, which was performed March 15, 1875, as follows :

Third Operation.—The fold of skin in the axilla being gathered up between the thumb and fingers, was divided across at two points two inches apart, and the incisions extended through the sound skin a short distance on either side. Between the elbow and shoulder the cicatricial band was also divided across at two points. After relaxation thus obtained, the extension of the limb could be carried to the utmost degree without any remaining resistance. The same treatment of the newly-made wound-surfaces and the same management of the limb itself were continued as after the previous operations.

April 28th.—Cicatrization being again completed with a satisfactory result, an expedient was resorted to which, it was hoped, would permanently resist and eventually overcome entirely the tendency to recontraction on the part of the axillary fold. It consisted of two tubular rings made of stout linen and stuffed with bran, and adapted to be worn high up over the shoulders. These rings, when applied

and drawn toward each other upon the back by
means of an elastic strap and buckle, exerted a constant pressure upon the axillary folds in front, and

FIG. 86.

resisted the contraction and consequent elevation of
the fold. A trough-like splint, incasing the arm and
forearm, and maintaining the elbow in a straight
position, was also worn constantly.

The final result of the treatment at the date of his discharge from the hospital (August 17, 1875), was as follows: The limb was capable of the utmost freedom of motion, and, when raised to a vertical position with the elbow and wrist fully extended, there remained no longer any resistance on the part of the fold in the axilla, or along the line of the original cicatricial band that passed down over the bend of the elbow. The shoulder-rings, which are still worn, had effectually prevented any contraction or elevation of the fold in the axilla, and had maintained it in a soft and pliable condition.

Fig. 86, from a photograph taken in July, shows this result.

THE END.

www.ingramcontent.com/pod-product-compliance
Lightning Source LLC
Chambersburg PA
CBHW021528210326
41599CB00012B/1426